RETURNED BY 10.15 AM
DATE STAMPED BELOW

17. MAY 96	21. FEB 97
21. MAY 96	25. FEB 97
22. MAY 96	26. FEB 97
22. MAY 96	27. FEB 97
24. MAY 96	28. FEB 97
	03. MAR 97
2 8 MAY 1996	04. MAR 97
	10. MAR 97
2 0 SEP 1996	11 MAR 1997

FINES PAYABLE FOR LATE RETURN:
80P FOR THE FIRST HOUR OR PART OF AN HOUR
50P FOR EACH SUBSEQUENT HOUR
OR PART OF AN HOUR

last revision August 1995   20113

ANTIBODY TECHNOLOGY

UNOTT

The INTRODUCTION TO BIOTECHNIQUES series

Editors:

J.M. Graham Merseyside Innovation Centre, 131 Mount Pleasant, Liverpool L3 5TF

D. Billington School of Biomolecular Sciences, Liverpool John Moores University, Byrom Street, Liverpool L3 3AF

Series adviser:

P.M. Gilmartin Centre for Plant Biochemistry and Biotechnology, University of Leeds, Leeds LS2 9JT

CENTRIFUGATION
RADIOISOTOPES
LIGHT MICROSCOPY
ANIMAL CELL CULTURE
GEL ELECTROPHORESIS: PROTEINS
PCR
MICROBIAL CULTURE
ANTIBODY TECHNOLOGY

Forthcoming titles

GENE TECHNOLOGY
LIPID ANALYSIS
GEL ELECTROPHORESIS: NUCLEIC ACIDS
PLANT CELL CULTURE
LIGHT SPECTROSCOPY
MEMBRANE ANALYSIS

ANTIBODY TECHNOLOGY

Eryl Liddell
School of Molecular and Medical Biosciences, Museum Avenue,
PO Box 911, Cardiff CF1 3US, UK

Ian Weeks
Department of Medical Biochemistry, University of Wales College of
Medicine, Heath Park, Cardiff CF4 4XN, UK

βIOS
SCIENTIFIC
PUBLISHERS

© BIOS Scientific Publishers Limited, 1995

First published 1995

A CIP catalogue record for this book is available from the British Library.

ISBN 1 872748 87 2

BIOS Scientific Publishers Ltd
9 Newtec Place, Magdalen Road, Oxford OX4 1RE, UK
Tel. +44 (0) 1865 726286. Fax +44 (0) 1865 246823

DISTRIBUTORS

Australia and New Zealand
  DA Information Services
  648 Whitehorse Road, Mitcham
  Victoria 3132

India
  Viva Books Private Limited
  4325/3 Ansari Road
  Daryaganj
  New Delhi 110002

Singapore and South East Asia
  Toppan Company (S) PTE Ltd
  38 Liu Fang Road, Jurong
  Singapore 2262

USA and Canada
  Books International Inc.
  PO Box 605, Herndon, VA 22070

Typeset by Chandos Electronic Publishing, Stanton Harcourt, UK.
Printed by Information Press Ltd, Oxford, UK.

100065 1117

Contents

Abbreviations

8-AG	8-azaguanine
ADCC	antibody-dependent cell-mediated cytotoxicity
ADEPT	antibody-directed enzyme prodrug therapy
AP	alkaline phosphatase
APAAP	alkaline phosphatase–anti-alkaline phosphatase
C	constant
CDR	complementarity determining regions
CEA	carcinoembryonic antigen
CG	chorionic gonadotropin
CHO	Chinese hamster ovary
DAB	3.3'-diaminobenzidine
dAb	single-domain antibody
EBV	Epstein–Barr virus
EF-2	elongation factor 2
ELISA	enzyme-linked immunosorbent assay
EMIT	enzyme-multiplied immunoassay technique
FACS	fluorescence-activated cell sorter
FCA	Freund's complete adjuvant
FIA	Freund's incomplete adjuvant
FITC	fluorescein isothiocyanate
FSH	follicle-stimulating hormone
GVDH	graft-versus-host disease
HAMA	human anti-mouse antibody
HAT	hypoxanthine–aminopterin–thymidine
HGPRT	hypoxanthine guanine phosphoribosyl transferase
HIV	human immunodeficiency virus
HRP	horseradish peroxidase
id	idiotypic
Ig	immunoglobulin
IL	interleukin
IRMA	immunoradiometric assay
IVIG	intravenous immunoglobulin
J	joining
LH	luteinizing hormone
MHC	major histocompatibility complex
MPO	microperoxidase
NK	natural killer
PAP	peroxidase–antiperoxidase
PCR	polymerase chain reaction
PEG	polyethylene glycol
RIA	radioimmunoassay

RSV	respiratory syncytial virus
RTA	ricin A chain
SCID	severe combined immunodeficiency
SDS	sodium dodecyl sulfate
6-TG	6-thioguanine
TK	thymidine kinase
TSH	thyroid-stimulating hormone
UV	ultraviolet
V	variable
VLS	vascular leak syndrome

Preface

This book deals with the methodology and applications of antibody-based techniques in research, diagnostics and therapeutics. Part 1 presents the basic principles and methods of both polyclonal and monoclonal antibody formation, including the latest genetic manipulation techniques. Part 2 focuses on applications of these antibodies in various diagnostic methods such as immunoassays and immunolocalization, together with a description of their use in a therapeutic context. Other applications are also dealt with, such as reaction catalysis and immunopurification.

It is written for both students and more experienced scientists who are new to antibody technology and for those who require an update on the latest developments. It is not meant to be a practical manual for antibody production or applications but, rather, it is hoped that enough information is provided to enable the reader to have an overview of the diverse uses of antibodies. Those who wish to pursue the subject further are directed to Appendix C, end-of-chapter references, and the list of suppliers' addresses and databases (Appendix B).

We have tried to keep references to original papers to a minimum, referring mainly to general immunology textbooks and chapters for established information. However, where these are not yet available for new antibody technology and applications, original papers have been cited.

We wish to convey our thanks to the Editors for their helpful advice and encouragement during the preparation of the manuscript.

Eryl Liddell
Ian Weeks

1 The Immune System

1.1 Introduction

The various systems of the mammalian organism are potentially at risk of infiltration by a wide variety of infectious micro-organisms. Most often, pathological damage is limited by the immune system. The immune system has two ways of overcoming such a challenge. The first of these arises from the innate immune system and the second from the adaptive immune system. The former is essentially a nonspecific first line of defence whereas the latter, as the name implies, is a mechanism specific to the immunological challenge. A further feature of the latter is a 'memory' effect which enables the immune system to react more rapidly and strongly should subsequent challenges be made by a given infectious agent.

The innate immune system comprises a variety of cells and humoral factors. The cells, namely phagocytes and natural killer (NK) cells, engulf and destroy 'foreign' particles. NK cells are also capable of binding to, and sacrificing, host cells that have exhibited cell surface changes as a result of viral infection or neoplasia. The actions of these cells are often mediated by soluble factors such as acute phase proteins including those of the complement cascade.

In certain situations, the invading micro-organism may not be effectively neutralized by the innate immune system, possibly because the micro-organism is not bound by the phagocyte either directly or via complement. In such circumstances, the mechanism of adaptive immunity is required to neutralize and eliminate the invading microbe. One requirement of this mechanism is the production of specific humoral factors capable of assisting phagocyte action which otherwise would not exist as part of the innate immune system. Such humoral factors are known as antibodies and are molecules produced by the B lymphocytes. Antibodies are the most important components of the immune system in the present context and are described in detail in Section 1.2.

B lymphocytes carry receptors on their surface capable of binding to surface molecules on the micro-organism. These receptors are similar in structure to the antibodies, which are ultimately secreted by the mature plasma cells derived from the B lymphocytes, in that they are glycoproteins known as immunoglobulins. The immune system possesses an enormous diversity of immunoglobulins capable of binding any large, 'foreign' invading molecule. Molecules capable of stimulating the immune response are known as antigens. The enormous number of possible antigens thus warrants great immunoglobulin (antibody) diversity.

1.2 Antibodies

Antibodies are glycoprotein molecules known as immunoglobulins. They are secreted by plasma cells derived from B lymphocytes which have been stimulated by binding of 'foreign' molecules to B-lymphocyte antigen receptors. The antibody has the same binding specificity as the antigen receptor.

Five classes of immunoglobulins exist in mammals and the characteristics of these are given in *Table 1.1*. The basic immunoglobulin unit consists of two heavy and two light polypeptide chains which are joined by disulfide bridges. The chains themselves also contain intra-chain disulfide bridges. Schematic structures of the basic immunoglobulin molecule are shown in *Figure 1.1*.

Each class of antibody possesses heavy chains characteristic of that class. Various subclasses show slight variation in the heavy chain structure within a given class. In all classes and subclasses the light chains can be one of two variants known as κ and λ.

TABLE 1.1:   *Classes of antibody molecules*

Immunoglobulin	Molecular weight	Approx. plasma concn (ng ml^{-1})
IgG1	146 000	9
IgG2	146 000	3
IgG3	170 000	1
IgG4	146 000	0.5
IgM	970 000	1.5
IgA1	160 000	3
IgA2	160 000	0.5
IgA (secretory)	385 000	0.05
IgD	184 000	0.03
IgE	188 000	5×10^{-5}

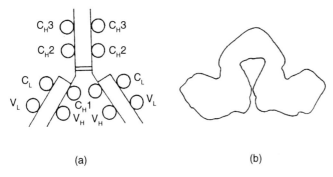

FIGURE 1.1: *Schematic representation of a typical IgG molecule showing (a) structure and (b) shape. C, constant; V, variable; H, heavy; L, light. Polypeptide loops are formed by disulfide bridges to yield the various domains represented here as circles. The two heavy chains are joined by disulfide bridges. The light chains are also joined to the heavy chains by disulfide bridges. Carbohydrate chains are not shown here.*

The molecules can be described as possessing both variable and constant regions, the latter being reasonably similar structurally within all antibodies within that class. The variable regions contain hypervariable regions which are responsible for antigen binding. The area of the hypervariable region which represents the actual binding site is known as the paratope and it is this region of the molecule which binds to the relevant antigenic determinant (epitope) on the antigen.

The different subclasses of immunoglobulin (Ig) are characterized particularly by differences in the number and position of inter-chain disulfide bridges. Immunoglobulin G (IgG) is the major immuno-globulin class in normal serum and is the major antibody of secondary immune responses, i.e. following a challenge subsequent to initial antigen exposure.

Immunoglobulin M (IgM) accounts for approximately 10% of the immunoglobulin pool, but is the major antibody of the primary immune response. It corresponds roughly to a pentameric form of the basic IgG molecule together with a so-called J (joining)-chain.

Immunoglobulin A (IgA) exists in circulating and secretory forms. In most mammals apart from man serum IgA exists predominantly as a dimer. IgA has two subclasses. Secretory IgA is dimeric consisting of two basic immunoglobulin molecules together with a J-chain and also a secretory component. IgA is the predominant immunoglobulin in seromucous secretions.

Immunoglobulin D (IgD) represents less than 1% of total immuno-globulins and its specific role remains unclear. It is however known to be present in relatively large amounts on the surface of circulating B lymphocytes.

Immunoglobulin E (IgE) is another trace circulating immunoglobulin but is found on the surface of basophils and mast cells. It is associated with acute hypersensitivity reactions such as allergies.

1.3 Antibody biosynthesis

It is not possible for each particular antibody specificity to be derived from a dedicated gene since there is simply not enough space in the genome to code for every conceivable antibody. Antibody diversity is thus generated in a number of other ways:

(i)   multiple germ line variable region genes;
(ii) recombination of variable region gene segments with other gene sequences;
(iii) inaccuracy of recombination;
(iv) somatic mutation;
(v)   combination of different heavy and light chains.

As can be seen from *Figure 1.1*, each immunoglobulin molecule possesses at least two variable regions, depending on the class of immunoglobulin. These variable regions in turn contain the paratope (i.e. that portion of the variable domain which binds to the epitope on the antigen). Both the light and heavy chains contain variable regions. The light chain also contains a constant domain whereas the heavy chain contains three constant domains.

Separate segments of DNA code for the variable (V) and constant (C) regions. In antibody-producing cells these segments are brought closer together. A further segment of DNA (J segment) is joined on to the V gene segment (*Figure 1.2*). A similar situation is seen in the case of heavy chains where a further gene segment (D segment) is joined on to the V and J segment genes (*Figure 1.2*). Thus diversity is generated by these recombinations and also by the fact that the position at which the various V, J and D segments combine may vary. In addition, it is common for further variation to be introduced due to point mutations in the V segment genes. Finally, additional diversity is generated due to the fact that almost any light chain may combine with any heavy chain.

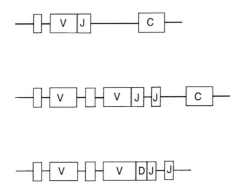

FIGURE 1.2: *Schematic representation of examples of B-lymphocyte DNA coding for antibody domains. (**Top**) λ light chain, (**middle**) κ light chain, (**bottom**) heavy chain variable domains. The sequences can be assembled as recombinations of many different V, C, J and D gene segments. Note that unlabeled boxes represent leader sequences.*

Immunoglobulin class and subclass is described by the heavy chain constant region. All the constant region genes are positioned downstream from the J segment genes. During class switching as part of an immune response, the variable region specificity is maintained. Thus, the initial IgM antibodies and subsequent IgG antibodies will have the same specificity but will merely be different isotypes. This relatively minor diversity is known as isotypic variation and is in contrast to the wide diversity of the binding site which is known as idiotypic variation. One further type of variation is due to intraspecies allelic variability and is known as allotypic variation.

When a B lymphocyte binds to the invading antigen the cell is stimulated and proliferates and matures. A given B lymphocyte carries only antigen receptors of a single specificity and produces only antibodies of that specificity. This is known as clonal selection. Proliferation thus leads to both antibody-producing cells and memory cells. Generally, many different B lymphocytes are stimulated, resulting in the production of many different antibodies. A population of such antibodies is known as a polyclonal population. The antibodies arising from a single line or clone of B lymphocytes are known as monoclonal antibodies.

Part of the antibody molecule binds to the specific antigen whilst another part is bound by receptors on phagocytes and may also activate complement. This ultimately results in the elimination of the micro-organism possessing that particular antigen. Antigen also causes clonal activation of T lymphocytes which differentiate into several subsets, namely helper, suppressor and cytotoxic T cells.

The innate and adaptive immune systems do not work in isolation, but interact at several levels. For example, activated T lymphocytes produce lymphokines which stimulate phagocytes. Furthermore, macrophages process antigen and present it together with accessory molecules necessary for T-cell activation.

Regulation of the immune response as a whole is clearly important. The antibody response is regulated in a number of ways. Following initial antibody formation, both antibody and circulating immune complexes give rise to suppression and/or augmentation of further antibody production. The specific effect is dependent on the class and subclass of the antibodies concerned. A further means of regulation is proposed by Jerne's network hypothesis. This hypothesis is based on the fact that the hypervariable regions of immunoglobulins can themselves act as antigens. The antibodies produced in this way are known as anti-idiotype antibodies since they bind to the idiotypic (id) region of the first antibody (*Figure 1.3*). Such anti-id antibodies may be directed to the binding site (paratope) or to regions adjacent to the binding site of the first antibody. In the former case, the anti-id antibodies are capable of inhibiting binding of antigen and represent an 'internal image' of the original antigen. Furthermore, the anti-id antibodies themselves are capable of eliciting antibody formation via their hypervariable regions. This network of antibodies is proposed to form the basis of a regulatory framework. However, much of the evidence for the presence of such a network is indirect and its specific regulatory mechanism remains to be established.

In summary, the mechanism of the aquired immune response involves the production of immunoglobulin molecules or antibodies. These molecules are capable of binding specifically to an antigen which represents a challenge to the host. Whilst the binding of antibodies to

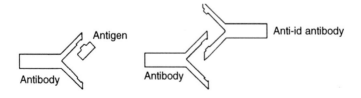

FIGURE 1.3: *(Left) Antibody–antigen binding. (Right) Antibody–anti-id binding. In cases where the anti-id antibody binds to the antigen binding domains of the first antibody, the anti-id binding site is said to correspond to the 'internal image' of the antigen.*

antigens is established *in vivo* it must also be recognized that such physical interactions can take place *in vitro* under the appropriate conditions. In this way it is possible to exploit the specific binding properties of antibodies for a wide variety of purposes.

2 Production of Polyclonal Antibodies

From the previous chapter, it can be seen that the mammalian organism is capable of producing immunoglobulin molecules (antibodies) in response to an enormous variety of intrusive chemical species which the animal's immune system perceives as being foreign and therefore potentially harmful. Such antibodies obviously fulfill a vital role *in vivo,* but are sufficiently robust to be used as therapeutic or diagnostic tools.

2.1 Manipulation of the immune response

In order to make full use of the possibilities offered by antibody reagents, it is necessary to be able to manipulate the immune response such that antibodies can be produced having the desired specificity and affinity for binding the antigen of interest. A prerequisite of this is that the immune system of the animal concerned must 'see' the antigen as foreign. For example, a rabbit will generally regard human proteins as foreign and mount an antibody-producing immune response. This however presupposes that the human protein is substantially different from any 'self' proteins present in the rabbit to which its immune system is tolerant.

In order to elicit an antibody-producing immune response, the putative antigen must be sufficiently foreign and sufficiently large. Small molecules, even if foreign to the animal, will not elicit an immune response. However, it is possible to 'trick' the immune system by derivatizing the small molecule in such a way that it is then perceived as a large, foreign molecule. In this case the small molecule is referred to as a hapten rather than an antigen. Derivatization of haptens to yield antigenic conjugates is discussed in Section 2.4.1.

A wide variety of animals have been used as host animals for antibody production. Mice and rats are primarily used for eliciting the initial polyclonal response necessary for the production of monoclonal antibodies (see Chapter 3). Guinea pigs and rabbits are mainly used in a laboratory situation for the production of polyclonal antibodies, whereas sheep and goats are the most widely used for large quantity antibody production.

2.2 Antibody production

The production of useful polyclonal antibodies is largely an empirical process. There are only a few variations of the basic immunization procedure which can increase the chances of obtaining useful material. Primarily, these variations involve route of delivery of immunogen, dose of immunogen, timing and number of immunizations.

The use of animals for antibody production is regulated by the Home Office and legislation set out in various relevant acts, particularly the Animals (Scientific Procedures) Act of 1986 [1]. This requires that both the particular project and the individuals involved must be licensed before the necessary work can be done. Furthermore, the project may only be undertaken in an establishment designated for animal work by the Home Office.

2.2.1 Mode of immunization

The most effective means of exposing the immune system to an antigen in order to yield a hyperimmune state is to present it as a sustained release preparation in conjunction with an agent known to cause general stimulation of the immune system. Both requirements can be satisfied by use of an adjuvant. The most commonly used adjuvant for immunization is known as Freund's complete adjuvant (FCA). This consists of a mineral oil base containing a suspension of killed mycobacteria. When an aqueous solution of the antigen is added to this, a biphasic system initially forms. Vigorous agitation of the mixture however leads to the formation of a thick emulsion. The production of a stable emulsion is often regarded as something of an art and most researchers have their own preferred methods. Several other vehicles have been reported but there appears to be no universal advantage in their use save where the method outlined above has failed to yield an appropriate response. Examples of such presentations are alum precipitates or loaded liposomes.

Effective immunization requires several administrations of immunogen. The procedure outlined above with FCA is normally regarded as the primary immunization. This gives rise to the so-called primary immune response which results in formation of a low concentration of IgM antibodies. Further challenge of the immune system results in class switching to the more useful IgG class of antibodies, which are also produced at much greater concentrations (high titer). This challenge is presented as the secondary immunization and in practical terms is performed in the same way as the primary immunization, except that the FCA is replaced by Freund's incomplete adjuvant (FIA). This essentially consists of FCA without the presence of the killed mycobacteria.

Further immunizations (booster immunizations) are carried out as necessary in an attempt to maximize antibody production and are used prior to harvesting a bleed from the animal. These immunizations are essentially further secondary immunizations in the case of polyclonal antibody production.

The site of immunization depends on the stage of the immunization regime and the animal being used. In mice, the immunogen is usually administered as a single subcutaneous bolus at the back of the neck. In rabbits, three or four sites along the flank are used and in sheep a similar number of sites are used on the rump. In the latter case, intradermal or intramuscular administrations have also been shown to be effective.

2.2.2 Dose of immunogen

It is accepted that the highest affinity antibodies (see Section 2.3.2) are produced by immunization with relatively small quantities of immunogen. Depending on the animal, a dose of 10–100 µg of immunogen is normally used. In a mouse, this would typically involve administration of 100 µl of emulsion or in the case of a sheep, up to 1 ml.

2.2.3 Timing of immunizations

The timing of the various immunizations varies depending on the animal. In mice, the secondary immunization is administered approximately 2 weeks following the primary immunization. In larger animals, this interval is substantially longer, possibly up to 2 months. Also in the larger animals, the booster immunization is administered some 2 weeks prior to harvesting. In mice, the booster immunization is optimally administered only a few days prior to splenectomy for fusion purposes (see Chapter 3).

The production of antibodies during the immunization regime is normally monitored using immunochemical assays. These will be discussed in Chapter 5. The mode of collection and test bleeds varies depending on species. Test bleeds are obtained most commonly by venepuncture. In mice the tail vein is commonly used, in rabbits the marginal ear vein is easily accessible and in sheep the jugular vein is most easily utilized. The latter case is also used for large volume collection in sheep; however, large volume collection from smaller animals often requires the use of cardiac puncture, which is invariably a terminal procedure carried out under anesthesia.

The blood collected from the immunized animal is normally allowed to clot to permit separation of the serum from the blood cells. This product is conventionally known as the antiserum and for long-term storage is usually kept frozen in aliquots to avoid repetitive freeze–thaw cycles.

2.3 Characterization of antibodies

The ultimate test of an antibody is whether it is usable for the technique for which it is intended. However, it is often necessary to achieve some level of qualitative and quantitative information prior to its final utility or as a preliminary means of assessing the success of an ongoing immunization regime by the characterization of test bleeds.

A variety of parameters are often assessed. The following are of particular importance: titer, affinity, cross-reactivity, immunoglobulin class (monoclonal antibodies only).

2.3.1 Titer

The definition of antibody titer is at best semi-quantitative. The term is most often used as a means of indicating the effective amount of specific antibody in an antiserum. The heterogeneity of polyclonal antibody populations precludes the ability to define antibody concentrations in absolute terms, and so the titer is defined as that dilution of antiserum which is required to bind a given quantity of antigen under defined conditions. There are several ways of assessing antibody titer using various immunochemical techniques. Since there is no standardized method, figures expressing antibody titer have little absolute meaning unless one has a feel for the method of assessment. The concept of antibody titer is however useful for the relative assessment of antisera when the method of determination is kept constant. In this way, it is a useful parameter for comparing bleeds from different animals or for comparing successive test bleeds. Whilst more commonly associated with polyclonal antisera, the determination of antibody titer is also often of use in monitoring the amount of monoclonal antibody in cell culture supernatants.

Probably the most basic means of assessing antibody titer in qualitative terms is by solution-phase immunoprecipitation. All antibodies are multivalent and in situations where the antigen has more than one epitope, it is possible to bring about the formation of insoluble immune complex lattices. The formation of such lattices is dependent on the relative concentrations of antigen and antibody. In cases of antigen or antibody excess, no such lattices are formed. However, at the so-called point of equivalence, a lattice forms and immunoprecipitation occurs (*Figure 2.1*). Thus, by adding increasing amounts of antigen to an antiserum, there comes a point at which turbidity is apparent, and this signifies the presence of substantial quantities of specific antibody capable of binding that particular antigen.

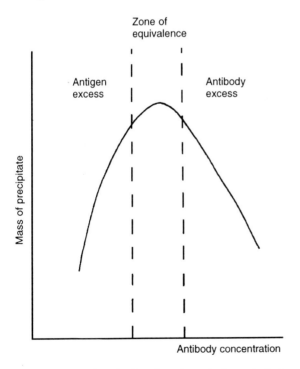

FIGURE 2.1: *Graph showing the relationship between the mass of immunoprecipitate and the amount of titrated antibody relative to a fixed amount of antigen.*

Semi-quantitative methods of antibody determination are offered by performing immunoprecipitation in semi-solid matrices such as agar or polyacrylamide gels. Here, antigen and antibody solutions are spatially separated by applying them to wells formed in a gel slab and allowing them to migrate towards each other by either passive diffusion or the effect of an electric field. Such methods are termed immunodiffusion and immunoelectrophoresis, respectively. There are numerous variations in performing these basic techniques, but generally assessment is by means of relative measurements of migration of the precipitin band from the origin or by densitometric assessment of the immunoprecipitate. Only relatively large masses of precipitate can be visualized directly by eye, but the sensitivity of the techniques can be enhanced by the use of gel staining (e.g. using Coomassie blue) to highlight the precipitin bands. However, in all the above cases only microgram to milligram quantities can be observed.

More discerning antibody assessment requires the use of immuno-assay techniques in their various forms. The reader is directed to Chapter 5 for a more detailed treatment of such techniques; however, the relevant aspects in the present context are dealt with here.

The most common means of determining antibody titer using immunoassay relies on the availability of labeled antigen (classically radiolabeled antigen) in order to assess the degree of formation of immune complex and hence the presence of antibody. If antibody is present, then clearly immune complexes will form with the labeled antigen. When the amount of added labeled antigen is kept constant, the amount of radioactivity taken up in the immune complexes reflects the amount of antibody present. However, it can be seen that, in order to quantify the activity of the labeled antigen in the immune complex, it is necessary first to separate immune-complexed labeled antigen from any labeled antigen not bound by antibody (*Figure 2.2*).

There are numerous methods of separation of antibody-bound and free antigen in an immunochemical mixture. Such methods usually resolve the mixture on the basis of physico-chemical differences. In

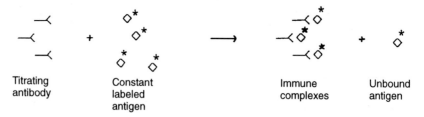

| Titrating antibody | Constant labeled antigen | Immune complexes | Unbound antigen |

FIGURE 2.2: *Schematic representation of antibody titration.*

situations where the antigen or its radiolabeled analog is a low molecular weight entity, any nonantibody-bound antigen can be separated by adsorption to charcoal. Antibody-bound antigen is not adsorbed and separation of the mixture is by means of filtration or centrifugation. The amount of activity measured in either the solid or liquid phases is thus related to the degree of immune complex formation. This method of separation has been widely used for small molecule-based systems such as steroids and small peptides.

In situations where the antigen is not adsorbed by charcoal it is possible to separate the immune complex on the basis of molecular weight/charge. In this case, a common method is to make use of selective precipitation of the immune complexes by polyethylene glycol followed by physical separation of the solid and liquid phases as before. A variation of this method involves formation and precipitation of a secondary immune complex. This is achieved by adding to the initial mixture a second antibody capable of binding to the first antibody, such that much larger immune complexes are formed. For example, if the first antibody was produced in a rabbit, then the second antibody might be an anti-rabbit immunoglobulin antibody produced in a sheep. Reliable precipitation in these cases is usually ensured by the presence of a dilute solution (2–3%) of polyethylene glycol.

Modern methods of separation are based on the use of pre-formed solid-phase reagents, these being most convenient for routine immunoassay purposes. A more detailed description of such materials will be given in Chapter 5, but essentially these consist of reagents in which either an antigen or an antibody is physically or chemically coupled to a solid matrix such as cellulose particles or plastic beads. This permits physical separation of the solid phase component of the subsequent reaction mixture. Thus, the titer of an antiserum produced in a rabbit could easily be estimated by mixing various dilutions of that antiserum with a constant amount of radiolabeled antigen. Following an initial incubation period, a suspension of particles consisting of cellulose-coupled sheep (anti-rabbit Ig) antibodies could be introduced with a further incubation. Separation of the solid-phase component followed by quantitation of radioactivity would thus yield a measure of the amount of specific antibody present in the antiserum under test (*Figure 2.3*).

2.3.2 Affinity

The strength of binding of an antibody to its corresponding antigen is described by the affinity constant. This parameter is actually a thermodynamic constant representing the equilibrium constant of the

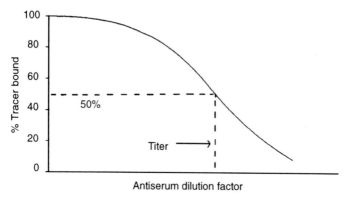

FIGURE 2.3: *Example of an antibody titration curve used for estimating the dilution of antiserum required to bind a given amount of labeled antigen.*

reaction between antigen (Ag) and antibody (Ab) (Equations 2.1–2.3).

$$Ag + Ab \underset{k_d}{\overset{k_a}{\rightleftharpoons}} AgAb \qquad (2.1)$$

$$K = [AgAb]/[Ag][Ab] \qquad (2.2)$$

$$K = k_a/k_d \qquad (2.3)$$

where k_a is the association rate constant, k_d is the dissociation rate constant, and K is the affinity constant.

In practical terms the affinity constant is established empirically by setting up an equilibrium reaction between the antigen and antibody and establishing the relative quantities of antibody-bound and free antigen. This is most commonly achieved by setting up a competitive binding immunoassay with various known concentrations of competing antigen relative to a fixed concentration of labeled antigen.

In order to calculate the affinity constant, it is necessary to perform a Scatchard transform [2] of the raw data so as to yield a plot of the ratio of antibody-bound/free antigen versus the concentration of bound antigen in moles per liter (*Figures 2.4* and *2.5*).

The use of antibodies for immunoassay purposes or for therapeutic purposes requires that the antibody has the highest possible affinity ($K > 10^{10}$ l mol^{-1}); in other words, it binds very strongly to its corresponding antigen and does not readily dissociate. By contrast, antibodies used for preparative immunoextraction, where the antigen

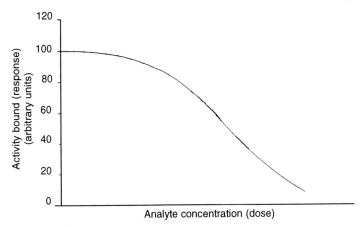

FIGURE 2.4: *Example of a typical dose–response relationship for a classical radioimmunoassay in which immune complex activity (labeled antigen bound to antibody) is represented as a function of the added dose of analyte antigen necessary to cause displacement of the labeled antigen from the antibody.*

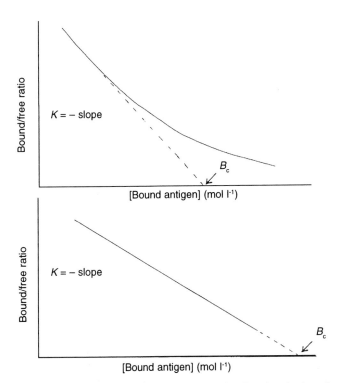

FIGURE 2.5: *Typical Scatchard plots for (**top**) polyclonal antiserum and (**bottom**) monoclonal antibody.* K *is the affinity constant and* B_c *the binding capacity.*

is to be recovered from the immune complex, are required to be of somewhat lower affinity ($K < 10^{10}$ 1 mol^{-1}) so that the chaotropic conditions used for dissociation are not so severe as to cause denaturation of the material to be recovered.

Another, less widely used, parameter obtained from the Scatchard analysis is the binding capacity, B_c, which is given by the point of intersection of the extrapolated plot with the [Bound] axis. The binding capacity represents the effective concentration of antibody binding sites in moles per liter.

Like the determination of antibody titer, the calculation of K and B_c may yield slightly different results depending on the conditions used to obtain the raw data, e.g. pH and ionic strength of buffers. Thus, for this reason, the method is best suited for comparisons of antibodies within a panel being investigated simultaneously where the immunochemical reactions are carried out in a reproducible manner. Absolute figures for antibody titer are thus not always a reliable parameter for accurate comparisons of antibodies.

In deriving Scatchard parameters it is important that the raw data are obtained as accurately as possible. This is normally best achieved by performing a competitive binding immunoassay over the full range of effective competition for labeled antigen. This would normally require the determination of binding at 10 different concentrations of competing antigen, each determined at least in triplicate.

The Scatchard plot is seen as a line or curve of negative slope. A monoclonal antibody would normally be expected to yield a straight line whose slope corresponds to $-K$ where K is the affinity constant in liters per mole. A curvilinear plot in such a case raises doubts as to whether the antibody actually is monoclonal or perhaps indicates that the affinity of the antibody for the labeled antigen is different from that of the competing, natural antigen. Polyclonal antisera normally yield curved plots which represent the fact that such antisera contain many antibodies having a wide range of affinities. Often, the slope representing the affinity constant of the highest affinity sub-population is quoted, but it is perhaps more representative to quote the range or average affinity of the antibodies in the antiserum.

2.3.3 Cross-reactivity

In most situations it is desirable that a given antibody be specific for a particular antigen and not react with other structurally related antigens. In reality this is not universally so and it is therefore important that potential cross-reactions be investigated. This is easily done, particularly in a competitive binding situation where possible cross-reacting species can be identified by their ability to compete with the labeled antigen for binding to the antibody as compared with the conventional competition by the 'authentic' antigen (*Figure 2.6*).

2.3.4 Immunoglobulin class

By its nature, a polyclonal antiserum will contain a plethora of antibodies differing in terms of specificity, affinity and class. However, a monoclonal antibody population will contain antibodies all of which are identical in every respect. In this latter situation it is often useful to know the class and subclass of the immunoglobulin molecule. This is very easily determined by using a range of anti-immunoglobulin reagents of known reactivity. Such reagents are widely available commercially and are themselves well-characterized antibodies. These reagents are highly specific for particular classes and subclasses of immunoglobulins. Thus, the monoclonal antibody of interest is incubated in turn with a panel of reagent antibodies each of which is specific for IgM, IgG1, IgG2a, IgG2b etc. Theoretically, the monoclonal antibody under study should react with only one of these reagents to form immune complexes, which immediately identifies its class and subclass.

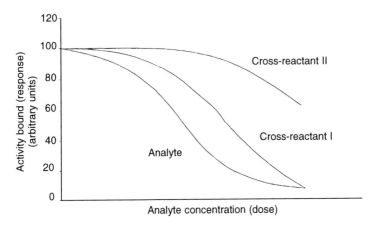

FIGURE 2.6: *Typical antibody cross-reactivity as shown by a series of dose–response curves for the authentic analyte antigen and two related species. The affinity of the antibody for the analyte is greater than the affinities for cross-reactants I and II; however, the presence of I or II in appreciable amounts in the sample would result in inaccuracy in the quantitation of the analyte.*

2.4 Chemical manipulations

At some stage during the production and use of antibodies it is probable that chemical manipulation will be necessary to modify various biochemical reagents. In the present context, it is necessary to consider the methods by which the following are produced:

(i)   hapten–carrier conjugates;
(ii) solid-phase reagents;
(iii) labeled reagents.

2.4.1 Hapten–carrier protein conjugates

It is often necessary to conjugate a small organic moiety to a larger, more complex molecule. This is particularly true for the production of small molecule immunogens, since the hapten in isolation will not elicit an immune response. Additionally, such conjugates are often useful as reagents in analytical or preparative systems where the nature of the isolated small molecule does not lend itself to simple production of solid-phase or labeled derivatives. In such cases, the protein portion of a suitable conjugate derivative is often better suited.

The chemical reaction that can be employed for the preparation of hapten–protein conjugates is constrained by the conditions which proteins will tolerate [3–5]. Thus, it is clear that reactions requiring extremes of temperature or pH, or the presence of high concentrations of organic solvents cannot be employed. Furthermore, the choice of chemical reaction is restricted by the nature of the hapten itself in terms of the functional groups which are available for reaction. Often, the hapten against which antibody reactivity is required does not possess appropriate functional groups and an analog must be employed which can undergo a suitable reaction. It is obviously important that the structure of the analog does not deviate too markedly from that of the antigen otherwise the requisite antibody reactivity will not be achieved.

The approach described above is exemplified in *Figure 2.7*. Here it can be seen that progesterone does not possess a useful functional group that can be used in a conjugation reaction with a carrier protein. However, a hydroxylated analog can form the basis of a synthetic precursor [6]. A further aspect of this strategy involves derivatization of the hydroxyl group to allow the subsequent analog to react under

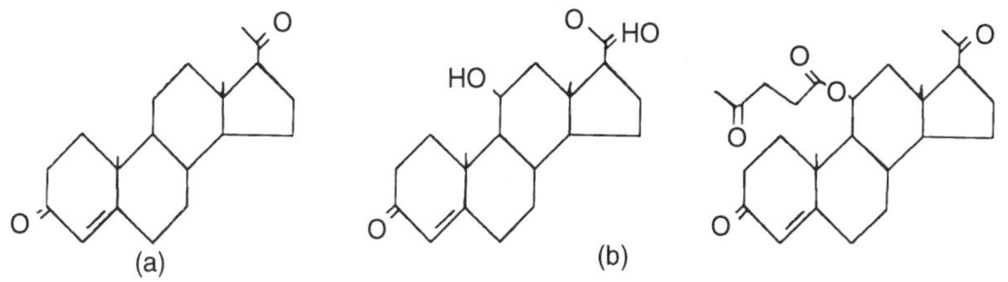

FIGURE 2.7: *(a) Progesterone and (b) two examples of analogs used for the preparation of conjugates for use as immunogens or immunoassay reagents.*

mild conditions with a carrier protein and also introduce a spacer between the hapten and the protein, thus overcoming possible steric hindrance to its ultimate use. In this example, the hydroxy-progesterone is converted to its hemisuccinate using succinic anhydride. The resulting carboxyl group can be coupled to proteins in a variety of ways.

An important consideration here is that the nature and position of the linkage with respect to the hapten analog can significantly affect the specificity of the final antibody in terms of its recognition of the natural hapten. It is particularly important in the case of polyclonal antiserum production that where a hapten–protein conjugate is being used in a downstream application of the antibody, the natures of the linking group and carrier protein are different to those used to produce the original immunogen. Failure to observe this results in aberrant cross-reactivities between bridge and/or protein epitopes. Even if such precautions are taken, it is frequently found that the antibodies have higher affinities for haptens modified at a point corresponding to a site used for original immunogen production, than for the natural hapten. Such disparate affinities can frequently be problematical. In some situations, it is often necessary to use downstream conjugates which differ in terms of carrier protein, nature of bridge (heterologous bridge) and position of bridge (heterologous site) [7]. *Table 2.1* shows a variety of ways in which various hapten functional groups can be conjugated to carrier proteins.

2.4.2 Solid-phase reagents

One of the major advances in the use of antibodies for both diagnostic and preparative purposes has been the advent of solid-phase technology. This encompasses the preparation of immunoassay reagents and insoluble supports for immunoaffinity purification matrices.

TABLE 2.1: *Examples of conjugation methods*

Hapten functional group	Coupling reagent/procedure	Protein functional group
General reagents		
COOH	Mixed anhydride	NH_2
	Carbodiimide	NH_2
	Via *N*-hydroxysuccinimide	NH_2
	Via carbonyldiimidazole	NH_2
NH_2 (aromatic)	Diazonium salt	Histidine
		Tyrosine
		Tryptophan
	Via isocyanate	NH_2
NH_2 (aliphatic)	Carbodiimide	COOH
	Diisocyanate	NH_2
	Glutaraldehyde	NH_2
OH	Via succinic anhydride etc. to yield COOH, thence as above	
OH (vicinal)	Via dialdehydes (periodate oxidation)	NH_2
C=O	*O*-(carboxymethyl)hydroxylamine to yield COOH, thence as above	
CHO	Schiff base formation	NH_2
Dedicated homobifunctional reagents		
NH_2	Difluorodinitrobenzene	NH_2
	Bis(imidoesters), bis(isothiocyanates), bis(succinimidyl esters)	NH_2
Heterobifunctional reagents		
NH_2/SH	Maleimide/succinimidyl ester	NH_2/SH
	Photoaffinity cross-linkers	Various

The range of solid-phase supports is diverse but can be crudely divided into 'macro' and 'micro' solid-phases. The range of possibilities is seen especially in immunoassay systems, where many techniques require the efficient separation of immune complexes prior to quantitation of bound antigen. Typical macro solid-phases are antibody-coated plastic tubes, microtiter plate wells, 'dip-sticks' and 6 mm plastic beads. Micro solid-phases include antibody-coated Sepharose, cellulose and magnetizable particles. These systems differ in their capacity and ease of separation (*Table 2.2*). Antibodies can be coupled to these solid-phases by either active or passive processes. The former often utilize chemical reactions similar to those described

TABLE 2.2:   *Solid phase matrices*

Microparticle	Macrosurface
Cellulose, glass, cross-linked dextran, paramagnetic particles	Coated tube, bead, well, dipstick
Advantages Rapid kinetics Easily controlled dosage High capacity	Ease of use
Disadvantages Complex separation	Slow kinetics Difficult to control dosage Batch-to-batch inconsistency

above for hapten–protein conjugation. The latter rely on the observation that, in appropriate buffers, antibodies will be adsorbed with high affinity on to plastic surfaces by a combination of hydrophobic and ionic interactions.

2.4.3 Labeled reagents

Many applications of antibodies require that they have a means of being visualized or quantified. This entails the introduction of some form of 'reporter' group into the antibody. A wide variety of reporter groups have been described. The most widely used labels are radioisotopes, fluorophores, latex particles, colloidal gold, enzymes and chemiluminescent molecules. Many others have been described but have not been used extensively. Again, the labeling procedure involves either active or passive techniques in much the same way as does the production of solid-phase reagents. Since the label used is substantially dependent on the application, more detailed descriptions of these various techniques will be deferred until Chapter 5.

References

1. *Guidance on the Operation of the Animals (Scientific Procedures) Act 1986.* HMSO Publications Centre, PO Box 276, London SW8 5DT.
2. Scatchard, D. (1949) *NY Acad. Sci.,* **51,** 660.
3. Parker, C.W. (1976) *Radioimmunoassay of Biologically Active Compounds.* Prentice-Hall, Englewood Cliffs, NJ.

4. Makela, O. and Seppala, H. (1986) in *Handbook of Experimental Immunology*, Vol. 1 (D.M. Weir, ed.). Blackwell, Oxford.

5. Erlanger, B.F. (1980) *Methods Enzymol.,* **70,** 85.

6. Kohen, F., Bauminger, S. and Lindner, H.R. (1975) in *Steroid Immunoassay* (E.H.D. Cameron, S.G. Hillier and K. Griffiths, eds). Alpha Omega, Cardiff, p. 11.

7. Corrie, J.E.T. (1983) in *Immunoassays for Clinical Chemistry* (W.M. Hunter and J.E.T. Corrie, eds). Churchill Livingstone, Edinburgh, p. 353.

3 Production of Monoclonal Antibodies

3.1 Introduction

Monoclonal antibodies are the secreted products of single clones of hybridoma cells. Hybridoma cells are created by fusing immune lymphocytes with myeloma cells such that the resulting hybrids possess both the antibody-secreting properties of the parent lymphocytes and the long-term survival potential of the parent myeloma in tissue culture. Through a series of selection procedures described in this chapter, single antibody-secreting hybridoma cells can be isolated in individual tissue culture wells from which large colonies (monoclones) can develop through mitotic division, each secreting identical monoclonal antibodies.

All the monoclonal antibodies from a single clone will bind to a single type of epitope or antigenic determinant on an antigen, rather than a range of different epitopes as would a polyclonal antiserum. This property can result in reagents of superb specificity, able to distinguish very slight differences between molecules or cells or micro-organisms. In addition, monoclonal antibodies can be made available in limitless quantities since the hybridoma cells can be grown in tissue culture virtually indefinitely and at industrial scales. There is also the added advantage that the cells can be frozen for storage, and recovered when required without the need for recharacterization as would be necessary for a new batch of polyclonal antiserum.

The vast majority of monoclonal antibodies are derived from mouse cells, although rat- and, more rarely, human-derived antibodies are also available. It is also possible to obtain antibodies derived from heterohybridomas which are products of parent cells from two of these species.

The production of monoclonal antibodies was first described in 1975 by Georges Kohler and Cesar Milstein in a paper entitled 'Continuous cultures of fused cells secreting antibody of predefined specificity' [1]. They were able to demonstrate for the first time that antibody-producing cells could be hybridized to immortal cells without compromising their ability to synthesize and secrete antibody. Thus, a continuous *in vitro* source of antibody of predetermined specificity could be obtained. This pioneering work has given rise to an explosion of new information in both fundamental biological research and in a wide spectrum of medical and diagnostic applications. It was recognized by the award of a Nobel prize in 1986.

This chapter describes the conventional methodology of monoclonal antibody production based on tissue culture, followed by various alternative procedures that have been applied to improve the final product.

3.2 Basic methodology

This section describes the most commonly used procedures for making monoclonal antibodies in the mouse system in order to provide the reader with sufficient understanding of what is involved in making them or what he/she may be buying. Each step in the process is illustrated in *Figure 3.1* and will be described in turn. Variations of these procedures will be considered in a later section. More detailed practical information for those who intend to make their own monoclonal antibodies can be found in other texts (see, for example, ref. 2).

3.2.1 Immunization

The production of immune lymphocytes essentially follows the same principles as the production of polyclonal antisera described in Chapter 2. However, there are some important considerations. Any strain of mouse or rat can be chosen for immunization but if the intention is to propagate antibody *in vivo* by ascitic fluid production, the strain should be the same as that from which the parent myeloma was derived, i.e. for mice that strain is usually Balb/c since the common myeloma cell lines are derived from this strain. Alternatively, ascitic fluid will have to be raised in F1 crosses of Balb/c and the immunized strain.

The purity of the immunogen is not quite so important as it is for polyclonal antiserum production since the selection of specific

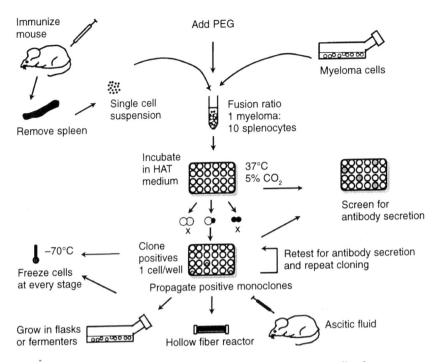

FIGURE 3.1: *Conventional hybridoma and monoclonal antibody production.*

antibody-secreting cell lines occurs after fusion and is dependent on the purity of the antigen used in the screening test. Indeed, it may be better to use an immunogen that has not been tampered with, if the requirement is to raise an antibody to a 'native' protein. The epitopes that monoclonal antibodies bind to are likely to be destroyed by alterations in the tertiary structure of the protein.

A conventional immunization protocol can be followed and the animals kept for up to a year before fusion. The serum antibody titer should be measured by a simple assay such as immunodiffusion or enzyme-linked immunosorbent assay (ELISA). Titers of the order of 1:1000 or more will give the best chance of recovering positive hybrids. In addition, animals should be boosted, preferably intravenously, 4 days before fusion. This will ensure that the relevant lymphocytes are at their peak of proliferation rather than antibody secretion.

The source of lymphocytes from rodents is normally the spleen (splenocytes), although lymph nodes have also been used. A fresh single cell suspension of the tissue, normally without further purification, is made immediately prior to fusion (see Section 3.2.4).

For human fusions, peripheral blood lymphocytes are used, mainly because of the ease of availability (see Section 3.4.1).

3.2.2 Tissue culture

A fully equipped tissue culture laboratory is essential for monoclonal antibody production. If possible, exclusive facilities should be made available, separate to those where primary tissue culture is carried out, in order to minimize the risk of cross-contamination. Essential equipment includes a laminar flow workstation, a CO_2 incubator, an inverted phase contrast microscope, a bench centrifuge and a freezer capable of achieving temperatures of $-70°C$ or less (liquid nitrogen storage facilities are preferable). A comprehensive range of disposable sterile plastic tubes and dishes, pipetting devices and culture media will also be needed, and are available from specialist suppliers, some of which are listed in Appendix B.

The successful practice of tissue culture does not require any special skills except for rigorous attention to aseptic technique. There are a number of good manuals (see, for example, ref. 3) and training courses are available, but there is no substitute for practice, preferably without antibiotics, before the more important cultures are begun.

3.2.3 Myelomas

During the 1960s, studies on the structure, biosynthesis and genetics of immunoglobulins stimulated the requirement for plentiful source material. Myeloma or plasmacytoma cells provided such material. They are neoplastic antibody-producing cells which can be generated easily in mice. Despite strenuous efforts, however, it proved extremely difficult to generate myelomas secreting antibody of known specificity, until Kohler and Milstein demonstrated that they could be fused with antibody-secreting lymphocytes [1].

Of the mouse myeloma cells adapted for survival in tissue culture, the most popular for hybridoma production are derived from the MOPC-21 myeloma. A variety of lines suitable for cell fusion (see *Table 3.1*) can be obtained from large international cell culture collections such as the American Type Culture Collection and the European Collection of Animal Cell Cultures (see Appendix B), from which stocks can be grown and stored frozen in individual laboratories. The quality of these lines is guaranteed and they are free of infection. It is not good policy to accept cell lines from other sources, however reputable, unless they also carry such guarantees.

TABLE 3.1: Commonly used rodent myeloma cell lines

Common name	Full name	Derivation	Ig secretion Heavy	Light	Original ref.
Mouse					
P3	P3/X63-Ag8	Balb/c myeloma MOPC21	IgG1	κ	[1]
NS-1	P3/NS-1/1-Ag4-1	P3	–	κ	[4]
Sp2	Sp2/0-Ag14	P3/spleen cell hybrid	–	–	[5]
653	P3X63-Ag8.653	P3	–	–	[6]
Rat					
Y3	Y3-Ag.l.2.3	LOU/C rat myeloma R210.RCY3	–	κ	[7]
YO	YB2/3.0 Ag20	Y3/AO rat spleen hybrid	–	–	[8, 9]
983	IR983F	LOU/C rat myeloma	–	–	[10]

These cell lines are available in the UK and Europe from the European Collection of Animal Cell Cultures, and in the USA from the American Type Culture Collection (see Appendix B).

There are two important characteristics of myeloma cell lines used in hybridoma production, which are described below: they should have an enzyme deficiency or unique chemical sensitivity or resistance to enable selection of fused from unfused cells (i) and, ideally, they should not secrete any immunoglobulin of their own (ii).

(i) Most myeloma cell lines are deficient in the enzyme hypoxanthine guanine phosphoribosyl transferase (HGPRT) although there are also others deficient in thymidine kinase (TK) and ouabain. HGPRT is essential for DNA and RNA synthesis. Lack of this enzyme prevents survival of myelomas in HAT medium (see Section 3.2.5), the culture medium in which post-fusion cell mixtures are grown, whereas fused cells are able to survive because they inherit the enzyme from the lymphocyte parent. The isolation of HGPRT⁻ myeloma cells is relatively easy because the enzyme is coded for on the single active X chromosome present in each cell. This means that only a single mutation is required to result in the loss of HGPRT. Selection of HGPRT⁻ cells is carried out in the presence of the toxic base analogs, 8-azaguanine (8-AG) and 6-thioguanine (6-TG), which are incorporated into DNA via HGPRT. It is not usually necessary to select for enzyme deficiency in myeloma cells since there is a range of useful HGPRT⁻ cell lines available commercially (see *Table 3.1*). The mechanisms by which these enzyme deficiencies are exploited in hybridoma selection are described more fully in Section 3.2.5.

(ii) The myeloma line P3, which Kohler and Milstein first used in 1975 [1], secreted immunoglobulin, so when fused with lymphocytes the hybrid cells secreted a mixture of heavy and light chains derived from both parents. This obviously reduces the chances of obtaining a hybridoma secreting antigen-specific antibody derived solely from the immunized lymphocyte parent. Soon after, myeloma lines became available that synthesized but did not secrete κ light chains. Thus, after fusion, hybrids could secrete myeloma-derived κ chains in addition to immunoglobulin chains from the lymphocyte parent. These myeloma lines, of which NS-1 became very popular, then gave way to fully 'nonsecretor' cell lines, which did not synthesize any immunoglobulin chains (e.g. 653, Sp2), and these are the lines of choice today.

Myeloma cells are very easy to grow in suspension tissue culture if certain characteristics are kept in mind. The growth medium to which they have been adapted will be recommended by the supplier (i.e. RPMI 1640 with 10% fetal calf serum) but they should not be grown continuously for long periods or at densities of more than 5×10^6 ml^{-1} in order to prevent overcrowding. When cells are put under stress, the likelihood of spontaneous mutations occurring increases. This could result, for example, in the line reverting back to being HGPRT⁺, and indeed there are some human myeloma lines which are particularly prone to this problem. Reliable stocks of cells should be kept at $-70°C$ or lower and initiated at frequent intervals as an insurance against spontaneous mutations, contamination problems or incubator breakdown.

3.2.4 Fusion

Although fusion is central to the whole process of hybridoma production, it does not require special equipment and takes approximately 30 min. Essentially, myeloma cells (in the log growth phase) and a single cell suspension of immune splenocytes (see Section 3.2.1) are mixed together in a sterile centrifuge tube (10^7–10^8 myeloma:10^8 splenocytes) and pelleted by gentle centrifugation. A solution of polyethylene glycol (PEG) (40–50%; M_r 1500–4000) is added slowly to the pellet over 1–2 min, then slowly diluted in culture medium, centrifuged and resuspended in fresh culture medium. The timing is crucial to avoid the potential toxic effects of PEG. The cell mixture is distributed between, typically, 3×96-well microtiter plates and allowed to grow in a 37°C/5% CO_2 incubator.

PEG fusion is not selective for cell types or antibody-secreting cells so, following fusion, a mixture of different cells will be present: unfused myeloma and spleen cells, hybrids of myeloma/myeloma, spleen/ spleen and myeloma/spleen, and also hybrids of more than two cells. The multiple hybrids will not survive for more than a few weeks and neither will the unfused spleen and spleen/spleen hybrids, because of their inability to grow in tissue culture. The unfused myeloma and myeloma/myeloma hybrids must be disposed of so that the only remaining cells are the myeloma/spleen hybrids. This is achieved by growing the cells in a selection medium such as HAT.

3.2.5 HAT selection

In order to understand the mechanism of HAT selection, one needs to appreciate that nucleic acid synthesis can follow one of two pathways: the *de novo* and the salvage pathways. All normal cells use the *de novo* pathway but if this is blocked they can bypass it and use the salvage pathway. The salvage pathway for purine synthesis uses the substrates hypoxanthine and guanine from which are formed inosinic-ribose phosphate and guanine acid-ribose phosphate, respectively, with the help of the enzyme HGPRT. Similarly, the salvage pathway for pyrimidine synthesis uses the substrate thymidine to form thymidine monophosphate with the help of the enzyme TK. Obviously, if both the *de novo* and salvage pathways are blocked, cells cannot survive.

HAT medium consists of the normal culture medium with three additives: hypoxanthine, aminopterin and thymidine. Aminopterin is an antibiotic which effectively blocks the *de novo* pathway, forcing all the cells to use the salvage pathway. Hypoxanthine and thymidine are the substrates utilized in the salvage pathway by HGPRT and TK, but if either of these enzymes is missing, as is the case with myeloma cells, purine or pyrimidine synthesis cannot occur and the cells will die. Hybrid cells, of enzyme-deficient myeloma and enzyme-positive lymphocytes, will be able to survive because they will have inherited the enzyme from the lymphocyte parent (see *Table 3.2*). After 5 days of growth in HAT medium, the only surviving cells in the fusion mixture will be myeloma/spleen hybrids.

Natural ouabain sensitivity is a property of use particularly in the selection of heterohybridomas of human/mouse origin. Human cell lines normally die in the presence of ouabain at 10^{-7} M, whereas rodent lines are resistant up to 10^{-3} M. Unfused human cells can therefore be selected against by including ouabain in the culture medium at concentrations of 10^{-6} M.

TABLE 3.2: *Post-fusion selection*

| Cell type | DNA synthesis | | Survival in HAT medium |
	Salvage pathway	*De novo* pathway	
Myeloma	HGPRT⁻	Aminopterin sensitive	Die (no DNA synthesis)
Spleen	HGPRT⁺	Aminopterin sensitive	Die (finite survival *in vitro*)
Myeloma/spleen hybrid	HGPRT⁺	Aminopterin sensitive	Live
Myeloma/myeloma hybrid	HGPRT⁻	Aminopterin sensitive	Die (no DNA synthesis)
Spleen/spleen hybrid	HGPRT⁺	Aminopterin sensitive	Die (finite survival *in vitro*)

3.2.6 Antibody screening tests

Approximately 2 weeks following fusion, it is necessary to determine which of the hundreds of mini cell cultures are producing the desired antibody, in order to reduce the numbers of cultures to a manageable and relevant number. Subsequently, antibody production will need to be screened frequently throughout the cloning stages. Theoretically, most immunoassays can be adapted for use in monoclonal antibody detection. In practice, however, selection of the most appropriate screening test requires careful consideration of the final proposed application of the antibody and the growth characteristics of the hybridoma cultures.

The screening test should be as similar as possible to the assay in which the antibody will be used; also, the antigen should be in the same form as it will be used in the final application. The reason for this is that an epitope to which a monoclonal antibody will bind is not necessarily composed of a continuous sequence of amino acids; it is likely to be made up of sequences from neighboring and overlapping chains. If this conformation is altered, the epitope may be destroyed and the antibody will not bind to the antigen. Thus, if a monoclonal antibody is selected on the basis of its binding to native protein, it may not react with denatured protein and vice versa. Therefore, a monoclonal antibody that works well in an ELISA may not recognize the same antigen after it has been exposed to sodium dodecyl sulfate electrophoresis and denatured. This is an important point to clarify when buying a monoclonal antibody.

It is preferable to design a screening test that is as specific as possible for the antigen of choice. Screening for general antibody production is not productive since most of the cell cultures will be producing antibody of some kind, so selection on the basis of the highest concentration of immunoglobulins will give misleading information unrelated to the levels of specific antibody secretion. Pure antigen is not always available, or if one is raising an antibody to distinguish between two different proteins or two different cell types, the specific antigen may not be known. In such situations, a secondary screening test for the major contaminant or alternative cell type may be of use.

The design of a useful screening test should also take into account the growth characteristics of the hybridoma cells. The cells grow very rapidly, with an exponential doubling time of 12–24 h, so there is a period of only about 3 days between the time when there is sufficient antibody secreted into the culture medium to measure and the time when the cells become so overcrowded that they begin to die. Maintenance of a healthy cell culture, with frequent replenishment of culture medium and space in which to grow, can conflict with the need to allow antibody to build up to measurable concentrations. Therefore, the sensitivity of the assay should be compatible with the need to pick up antibody secreted from a small colony of cells. Because of the need to identify the important cultures before they have outgrown the culture well, the screening test should also be quick to perform (24 h maximum) and able to accommodate several hundred test samples at a time. Thus, procedures that involve individual sample handling, centrifugation or observation of microscope slides, or lengthy preparation of antigen or labeled probes, would be counter-productive. If necessary, this type of test could be introduced in the early stages between cloning steps to support information provided by the larger screen.

An ideal assay format for screening monoclonal antibodies that can fulfill most of the criteria suggested above is an ELISA [11]. Hundreds of samples can be analyzed within a few hours, labeled reagents are easily and cheaply available, and relative concentrations can be determined visually, without the necessity for expensive equipment. A typical test format is shown in *Figure 3.2*.

3.2.7 Cloning

Cloning, in the context of hybridoma cells, simply means the dilution of a cell population, such that single cells can be isolated, which then grow into individual colonies. Thus, each cell within a colony is identical and so will secrete identical antibody. The most common method of achieving this is called *limiting dilution*. The cell mixture to be cloned is counted and diluted in a series of steps such that the theoretical number of cells per 100 µl (i.e. per well of a microtiter plate) is one. For practical purposes, many replicates of several

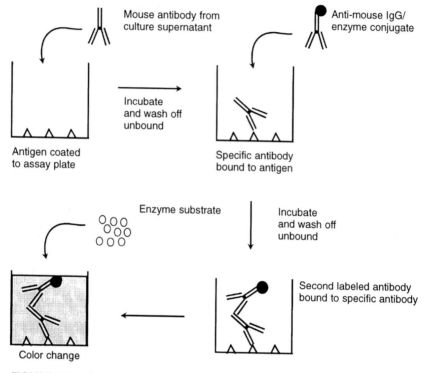

FIGURE 3.2: *Schematic diagram of simple ELISA format for antibody screening test: direct adsorption of pure antigen to assay plate. If pure antigen is not available, a specific polyclonal antibody could be coated to the assay plate before addition of the antigen mixture. Only pure antigen will then bind to the coating antibody.*

dilutions from 5 to 0.5 cells per well are grown, to increase the chances of obtaining monoclonal cell lines. These cultures must be supplemented with feeder cells or equivalent media supplements. Feeder cells are suspensions of thymocytes, peritoneal macrophages or splenocytes which secrete growth factors essential for clonal growth. Single clones in a well are selected by careful observation under the microscope, which is relatively easy if the culture has remained undisturbed and not dispersed (see *Figure 3.3*). The cloning efficiency will not be 100% and can be considerably less. It will depend on several factors such as cell numbers, initial cell viability, feeder cells or supplements. Within 2 weeks of cloning by this method the culture supernatant should be tested for secreted antibody. Those which are positive and also 'monoclonal' are transferred to larger culture wells and recloned (the remainder of the cells can be stored frozen). Recloning and antibody testing should be repeated until all the wells with growing cells are shown to secrete specific antibody. At this stage one can be reasonably sure that the cell line is monoclonal.

3.2.8 Storage (cryopreservation)

Once a monoclonal cell line has been established it is vital to secure the line by growing the cells up in flasks, harvesting and freezing. The freezing medium should contain 10% (v/v) dimethyl sulfoxide to prevent crystallization of water within the cells. Several aliquots of approximately 10^7 cells should be stored in several different freezers at $-70°C$ or in liquid nitrogen. It is also advisable to store a valuable line at one of the cell culture collections. The culture supernatant from these frozen cells can be used to characterize the antibody further, although higher titers can be obtained if cells are allowed to concentrate and die. However, concentrations of antibody obtained in this way rarely exceed 50 µg ml^{-1} and scaling up in stationary flasks is impractical.

3.2.9 Propagation of clones

There are two main alternatives based on *in vivo* or *in vitro* propagation. These are outlined as follows.

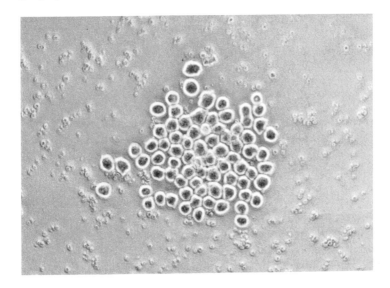

FIGURE 3.3:   *Photograph of monoclone. Reproduced from ref. 2 (Liddell and Cryer,* A Practical Guide to Monoclonal Antibodies). *Copyright (1991, John Wiley & Sons Ltd). Reprinted by permission of John Wiley & Sons Ltd.*

In vivo. The easiest way of obtaining very concentrated antibody (5–10 mg ml^{-1}) is to inject approximately 2 x 10^6 cells intra-peritoneally (i.p.) into histocompatible mice that have been pretreated with Pristane (0.5 ml i.p., 10–60 days before cell injection). Pristane encourages the growth of plasmacytomas and also depresses the normal immunological function of the recipient animal. Within 2 weeks of cell injection, the abdomen should swell with ascitic fluid which is rich in antibody (approx. 4 ml or up to 40 mg per mouse). The amount of antibody and fluid development is very dependent on the particular cell line. Some cells only induce solid tumors with very little fluid and, of course, no tumor or fluid will develop in mice that are not histocompatible. If the mouse immune lymphocyte donor was not Balb/c, then a F1 hybrid of that strain and Balb/c must be used to propagate ascitic fluid. The resulting fluid will be 'contaminated' with other mouse proteins including general immunoglobulins, but for many applications, at the typical working dilutions of 1:100 000 or more, these become insignificant.

One of the biggest problems with this method is that the Home Office Inspectors of the Animals Scientific Procedures Act (1986) in the UK consider it to be too traumatic for the animals and are imposing stricter guidelines for its use, particularly now that there are better alternatives in *in vitro* propagation. The technique is prohibited in Germany.

In vitro. Roller bottles and stirred flasks can be used to increase the concentration of antibody produced *in vitro* but the technique can be very labor intensive and requires large volumes of culture medium which need to be concentrated to produce antibody concentrations comparable with ascitic fluid. Fermenters or hollow fiber reactors are the preferred choice for larger scale production.

Fermenters have been used for many years to propagate bacteria but eukaryotic cells are more fragile and require more complex nutrition. One of the more successful types of fermenter for growing hybridoma cells is the air-lift fermenter. Cells are grown in suspension or on microcarriers in a vessel aerated from below in such a way that the gas bubbles gently mix the cells with minimal damage from sheer stresses that would be generated by conventional stirrers. Such fermenters range in size from 5 l for research production, but can be scaled up to a capacity for commercial quantities of 10^3–10^4 l, which can produce 40–500 mg of antibody per liter. The problem remains, however, of the expense of such large quantities of culture medium and the need to concentrate and purify the antibody.

Propagating hybridoma cells in hollow fiber reactors avoids this problem. Hollow fiber reactors are essentially bundles of fine porous fibers, rather like kidney dialysis cartridges, through which culture medium is circulated (see *Figure 3.4*). Cells are grown in a much smaller volume of medium in the extra-capillary space. Nutrients and waste products can pass freely between the two spaces but larger molecules such as antibodies remain in the cell compartment until harvested. Large densities of cells can be maintained in this system for months producing up to several hundred milligrams of antibody per day. The small volume cell growth compartment contains culture medium, including serum, but the large volume of circulating culture medium does not contain serum, thus costs are kept relatively low and the harvested antibody is already concentrated.

All the cell culture methods described so far require the use of serum in the culture medium which, because it is an animal product and ill defined, may interfere with some applications of the antibody. If this is the case, then the hybridoma can be grown in a fully defined serum-free medium. Hybridoma cell lines vary considerably in their capacity to grow well in serum-free media and they should be gradually adapted to the change by stepwise reduction of the serum percentage over several weeks. Some lines will never be able to grow without at least 2% (v/v) serum. However, there are numerous new products on the market to try, the latest information on which should be obtained from leading tissue culture suppliers (see Appendix B).

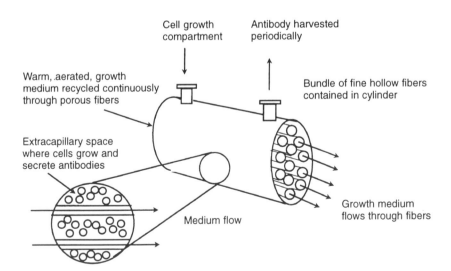

FIGURE 3.4: *Schematic diagram of hollow fiber reactor.*

3.3 Human hybridomas

Ever since mouse monoclonal antibodies were first made, the possibility of producing human antibodies for therapeutic purposes, in the same way, has been an exciting prospect. In practice, however, it has proved very difficult. Several factors have been blamed for this, including the instability of human myeloma cell lines, the difficulty in immunizing human volunteers against any but a small range of immunogens and the necessity of using peripheral blood lymphocytes rather than spleen in most situations. Nevertheless, human monoclonal antibodies have been raised by conventional hybridoma technology (see ref. 12 for review) and certain improvements and variations on the basic methodology (see below) have arisen out of these efforts. The best example of these is the application of recombinant DNA techniques to derive specific antibody fragments from immunoglobulin gene libraries, which has largely superseded the conventional approach and is dealt with in Chapter 4. The therapeutic applications of human antibodies derived by both methods are discussed in Chapter 7.

3.4 Variations on basic methodology of production

3.4.1 Alternative lymphocytes

The best rodent source of immune lymphocytes is the spleen which contains (in the mouse) approximately 10^8 cells, 50% of which are B cells. For most purposes, successful fusions can be achieved by using whole spleen suspensions without pre-selection of either B cells or those expressing specific antibodies. For obvious reasons, human immune spleens are not so readily available, although tonsil lymphocytes, an equally good source, can be used. In the majority of cases, peripheral blood lymphocytes have to be used. Despite their ready accessibility, blood is not an ideal source because there are not enough antigen-reactive specific B cells in the peripheral circulation at the appropriate stage of differentiation and proliferation, and there are more suppressor and cytotoxic cells. These problems can be alleviated by various cell enrichment and depletion strategies. 'Naive' lymphocytes can then be subjected to *in vitro* immunization (see Section 3.4.2) or immune lymphocytes can be transformed by Epstein–Barr virus (EBV).

EBV is a human B-lymphocytic herpes virus which is carried by an estimated 90% of the adult population. When incubated with B lymphocytes *in vitro*, it is capable of stimulating their growth and consequently antibody synthesis. In practice, supernatant from confluent cultures of an EBV-producing cell line, B95-8 (obtainable from the cell culture collections), is harvested and incubated with peripheral lymphocytes from recently immunized donors. This method has been used successfully on its own to produce antibodies (mainly of IgM at 20–50 μg ml^{-1}) but the lymphocytes can be very unstable due to the presence of memory T lymphocytes which are cytotoxic when stimulated *in vitro*. There can also be problems associated with regression of the proliferating foci of B lymphocytes after 3–4 weeks. Greater success has been obtained, particularly with human peripheral lymphocytes, by first stimulating B cells with EBV followed by fusion to mouse or human myelomas, or preferably mouse/human heteromyelomas [13].

3.4.2 *In vitro* immunization

In vitro immunization is the term used to describe antigen-specific activation of B cells in culture. The development of this method was prompted mainly by the difficulty of obtaining immune lymphocytes of a wide range of specificities from human donors. However, there can be additional advantages for rodent systems if *in vivo* immunization fails to produce the desired antibodies, i.e. when immunodominant antigens mask a response to less immunogenic epitopes or when trying to immunize against self-antigens. Also, hazardous substances such as tumor cells and toxins, which may be harmful to the whole animal, may be safe to use *in vitro*.

The objective of *in vitro* immunization is to parallel antigen-specific activation of B cells *in vivo* by providing soluble factors (lymphokines) and different cell types which are involved in the regulation of proliferation and differentiation of B cells. Many groups have had success by incubating rodent splenocytes or human peripheral blood B cells with mixtures of T cell-derived lymphokines (from mixed thymocyte cultures), activated thymoma-derived T cells or activated peritoneal macrophages, the adjuvant peptide N-acetyl muramyl-L-alanyl-D-isoglutamine and low concentrations of soluble or immobilized antigen for about 5 days before fusion (see ref. 14 for details of many different methods).

Antibodies have been obtained in this way from 'naive' lymphocytes and, also, treatment of pre-immunized lymphocytes prior to fusion is said to increase the proportion of specific hybrids. Additional advantages of this method are that 'immunization' can be much quicker (5 days) and much lower amounts of antigen (< 1 μg) are effective. The main disadvantages are that the method tends to produce mainly IgM antibodies (especially from naive lymphocytes) and high affinity antibodies are rare. However, the ability to manipulate and analyze precisely the components of the *in vitro* immunization mixture will lead to improved results and improved understanding of what is required to produce the desired antibody response.

3.4.3 Heteromyelomas and hybrid hybridomas

Heteromyelomas are derived from the fusion of two different myeloma cell lines from different species, for example, mouse and human. The aim was to provide a more stable immortal fusion partner for human lymphocytes. In order to select the hybrid it was necessary to introduce an additional selection mechanism to the usual HAT method. If one partner is HAT resistant, the other should be HAT sensitive and ouabain resistant, so hybrids are selected in medium containing HAT and ouabain. Once established, they have to be back selected for HAT resistance in order to be used in the usual way in fusions with lymphocytes. Alternatively, antibiotic resistance can be introduced to the human myeloma partner by transfection with a plasmid vector containing the resistant gene. The human myeloma cell line FU-266 was rendered resistant in this way to the antibiotic G-418 by transfection with the recombinant plasmid vector pSV2-neo[R] [15]. Heteromyelomas are then selected in medium containing antibiotic (which kills the mouse myelomas) and ouabain (which kills the human myelomas), the HAT sensitivity of both parents being retained.

Hybrid hybridomas are produced by the fusion of two hybridoma lines of different specificity or one hybridoma line and immune lymphocytes of different specificity. Again, selection of the hybrid requires different mechanisms for each partner such as those described above. The resultant antibody mixture contains a significant proportion of antibodies that are bispecific, i.e. each antigen-binding arm of the molecule recognizes different epitopes. For example, one arm may bind to a cell surface antigen and the other to a complement factor, so resulting in lysis of the target cell, or one arm may bind peroxidase and the other the target antigen, which would help to reduce nonspecific binding of secondary reagents in immunohistochemistry and immunoblotting. The applications of these constructs are discussed in more detail in Chapters 6 and 7.

3.4.4 Electrofusion

A method of fusing cells electrically was developed as an alternative to chemical fusion particularly for application to human cell fusion using very small numbers of cells [16]. Cells are drawn together by dielectrophoretic effects in a nonuniform alternating electric field of low field strength. When this cell suspension is then exposed to a high rectangular electric pulse of approximately 10 μsec duration, fusion of apposed cells occurs. A low ionic strength buffer is required in order to reduce temperature increases due to the electric field, and there is a tendency to produce hybrids of multiple cells. The inter-electrode gap is limited; thus, the maximum volume of cell suspension that can be used is in the range 10–200 μl (equivalent to about 10^6 cells).

Electroacoustic fusion can overcome some of these disadvantages since the initial cell alignment can be achieved by application of an ultrasound wave. In such cases, there are no limitations to the inter-electrode gap so physiological media and large volumes can be used. However, for most applications electrofusion is not a practical alternative to PEG fusion.

3.4.5 Alternative cloning methods

The fluorescence-activated cell sorter (FACS) provides a powerful tool for cloning [17]. The FACS machine is able to analyze individual cells on the basis of size, viability and fluorescence as a stream of single-cell-containing droplets is passed between laser beams. At the appropriate signal, individual cells can be deflected magnetically into individual culture wells. For hybridoma sorting, the cell mixture is first incubated with antigen-coated fluorescent beads (e.g. 0.9 μm diameter) which will bind to the relevant antibody expressed on the surface of the hybridoma cell. Unbound beads are washed away leaving an enhanced fluorescent signal on the desired hybridoma cells. FACS cloning is obviously an improvement on manual cloning for speed and efficiency but such a machine is beyond the means of many laboratories.

Cloning can also be carried out by diluting the hybridoma cell mixture and growing colonies on soft agar. If the antibody screening test is based on cell agglutination or lysis, these cells can be overlaid on the hybridoma colonies and those with areas of agglutination or lysis around them can be picked off for further processing. If the screening test required soluble antibody, however, the colonies would have to be picked off at random and grown in liquid culture before testing. In such circumstances, limiting dilution would be the method of choice.

3.5 Characterization of antibodies

The characterization of any antibody reagent is, of course, important (see Chapter 2), and the procedures are similar whether the antibody is polyclonal, monoclonal or engineered. Thorough testing of the specificity and titer of an antibody in the context in which it will be used, with relevant controls, must be carried out before it can be applied confidently. Unlike polyclonal sera, a monoclonal antibody can usually be used directly without further purification. When dealing with a monoclonal antibody, however, one should be aware that its unique specificity is for a particular epitope. i.e. not necessarily for the whole molecule. If that epitope is common to other molecules the antibody may also bind to apparently unrelated antigens. However, it is not always necessary to know precisely which epitope the antibody recognizes; for instance, a monoclonal antibody is often used to distinguish between two different cell types or molecules without knowledge of its unique specificity.

A monoclonal antibody will have been raised using perhaps only one type of screening test (e.g. ELISA) but, as discussed in Section 3.2.6, that antibody may not recognize the 'same' antigen in a different type of assay because of possible conformational changes in the antigen. Some monoclonal antibodies (and indeed hybridomas) are sensitive to being frozen such that their stability under different storage conditions may lead one to focus on a different hybridoma line of the same specificity.

Knowing the isotype (immunoglobulin class and subclass) of a monoclonal antibody is important in order to determine the most appropriate purification method, the appropriate 'second' antibody in an immunoassay and whether it would have the appropriate effector function such as complement fixation. The affinity of the antibody–antigen bond is useful to know since those of highest affinity are required for good immunoassays, whereas those of lower affinity are preferred for purification of antigen so that the bond can be broken easily without damaging the antigen.

3.6 Making or buying: relative merits

It is hoped that this chapter has given the reader a general understanding of what monoclonal antibodies are and how they are

TABLE 3.3: Basic requirements for monoclonal antibody production

Tissue culture facilities	Assay facilities
Equipment Laminar flow hood CO_2 Incubator Phase contrast microscope −70°C freezer/liquid N_2 container Bench centrifuge Multipipettor	Sufficient antigen for screening Hardware (i.e. ELISA plate reader) *Animal facilities* Relevant Government licences Immunogen
Consumable items Disposable tissue culture plasticware Culture medium	*Personnel* Preferably, full time, experienced (especially in tissue culture) for at least a year

made. Many monoclonal antibodies are now available commercially, in purified form or as native culture supernatant or ascitic fluid, and can be supplied with a variety of enzyme or fluorescent labels attached (see Appendix B). Buying antibodies can be very expensive. If this is a serious problem, an option worth investigating is to obtain the hybridoma cell line from a cell culture collection or the originator (often free for research purposes) and to make a batch of ascitic fluid oneself.

Obviously, if after checking the commercial databases (see Appendix B) and scientific literature, the particular antibody specificity required is not available, you must consider making your own monoclonal antibodies. Firstly, consider whether you actually need a monoclonal antibody. A well-characterized polyclonal antiserum can often be just as useful. Some polyclonal sera to synthetic peptides can be highly specific and of higher combined affinity than an equivalent monoclonal. Secondly, the time involved should not be under-estimated. In experienced hands and if all goes well, it usually takes a minimum of 4 months but could take more than a year, especially if dealing with a poor immunogen. Also the work involved is fairly intensive once tissue culture has begun. Thirdly, the costs can be substantial in labor, consumables and equipment if a laboratory is to be established for the first time. *Table 3.3* lists the basic requirements. Nevertheless, despite the frustrations of the production process, making one's own monoclonal antibodies can yield limitless quantities of useful, highly specific reagents which can also, if novel enough, be of commercial value.

References

1. Kohler, G. and Milstein, C. (1975) *Nature,* **256,** 495.
2. Liddell, J.E. and Cryer, A. (1991) *A Practical Guide to Monoclonal Antibodies.* John Wiley & Sons, Chichester.
3. Freshney, R.I. (ed.) (1992) *Animal Cell Culture: a Practical Approach.* IRL Press, Oxford.
4. Köhler, G. and Milstein, C. (1976) *Eur. J. Immunol.,* **6,** 511.
5. Shulman, M., Wilde, C.D. and Köhler, G. (1978) *Nature,* **276,** 269.
6. Kearney, J.F., Radbruch, A., Leisegang, B. and Rajewsky, K. (1979) *J. Immunol.,* **123,** 1548.
7. Galfre, G., Milstein, C. and Wright, B. (1979) *Nature,* **277,** 131.
8. Galfre, G., Cuello, A.C. and Milstein, C. (1981) in *Monoclonal Antibodies and Developments in Immunoasssay* (A. Albertini and R. Ekins, eds). Elsevier, Amsterdam, p. 159.
9. Kilmartin, J.V., Wright, B. and Milstein, C. (1982) *J. Cell Biol.* **193,** 576.
10. Bazin, H. (1982) in *Protides of the Biological Fluids* (H. Peeters, ed.). Pergamon Press, Oxford, p. 615.
11. Kemeny, D.M. (1991) *A Practical Guide to ELISA.* Pergamon Press, Oxford.
12. James, K. and Bell, G.T. (1987) *J. Immunol. Methods,* **100,** 5.
13. Roder, J.C., Cole, S.P.C. and Kozbor, D. (1986) *Methods Enzymol.,* **121,** 140.
14. Borrebaeck, C.A.K. (ed.) (1988) *In Vitro Immunisation in Hybridoma Technology. Progress in Biotechnology,* Vol. 5. Elsevier, Amsterdam.
15. Teng, N.N.H., Lam, K.S., Reira, F.C. and Kaplan, H.S. (1983) *Proc. Natl Acad. Sci. USA,* **80,** 7308.
16. Zimmerman, U. (1986) *Rev. Physiol. Biochem. Pharmacol.,* **105,** 175.
17. Parks, D.R., Bryan, V.M., Oi, V.T. and Herzenberg, L.A. (1979) *Proc. Natl Acad. Sci. USA,* **76,** 1962.

4 Antibody Production by Chemical and Genetic Engineering

4.1 Introduction

The antigen binding properties of a particular monoclonal antibody may be ideal in terms of its specificity and affinity, but its usefulness as a reagent may be limited by its isotype, such that certain effector functions cannot be harnessed *in vivo* or, alternatively, Fc receptors on the antigen interfere with specific reactions. Under such circumstances, it may be necessary to alter the antibody by fragmentation and change or remove the Fc region, or perhaps combine fragments of other antibody specificities to make bispecific reagents. The various constructs possible by chemical means are discussed in Section 4.2.

Considerable progress is being made in the isolation and manipulation of antibody genes. It is possible to create a far greater variety of antibody binding specificities from genes, the products of which can then be expressed in bacteria or mammalian cell systems. Genetic manipulation enables the process of acquiring the desired specificity to be quicker and more controlled than the random cell fusion process of conventional hybridoma production. Sequence information can be more readily obtained from genes than from proteins thus facilitating studies of the structure of the antibody binding site and the epitope. Indeed, once the desired antibody fragments have been obtained, their binding affinities and even specificities can be improved upon by genetic alteration. Different specificities can be obtained from the same gene library, and it is even possible to create antibody libraries without immunization. The different strategies employed to obtain these products are described in the latter part of this chapter.

4.2 Chemical modification

4.2.1 Antibody fragmentation: Fab and F(ab')$_2$

The structures of antibody molecules are relatively well characterized at both the protein and nucleic acid levels, with the binding sites for antigen, Fc receptors and complement being assigned to different parts (see *Figure 4.1*). The Fc part of an antibody is responsible for the effector functions of the molecule. For some applications, this can be a hindrance. For example, many cell surfaces possess Fc receptors to which the Fc region will bind regardless of the specificity of the antibody. In addition, the complement fixing ability of some Fc regions

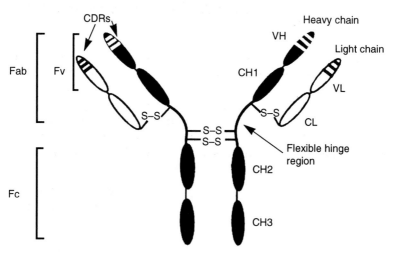

FIGURE 4.1: *Antibody structure (human IgG). This diagram illustrates the basic antibody structure of two identical heavy chains linked to two identical light chains and subdivided into domains of variable and constant amino acid sequences: V = variable; H = heavy; L = light. The variable regions of each chain contain three areas of hypervariable sequence known as complementarity determining regions (CDRs) which form the antigen binding site. The variable regions VH and VL together are known as Fv and VH, CH1, VL and CL together are known as Fab. The remainder of the antibody, Fc, joins the two Fab arms together via a flexible hinge region and disulfide bonds, and contains the effector functions of the antibody, i.e. those regions that bind to Fc receptors on target cells and to complement. Other antibody isotypes may vary by the number of domains in the Fc region, the number of disulfide bonds joining the chains and the number of these units joined together (i.e. IgM has five).*

might interfere in some assays. In other applications requiring cell penetration of the antibody or localization of an antigen, it may be an advantage to reduce the size of the antibody.

A long-established method of doing this is to cleave the Fc portion by proteolytic enzyme digestion (see *Figure 4.2* and ref. 1). Papain, a nonspecific thiol protease, splits the antibody at residue 224 in the hinge region, which results in two monovalent Fab fragments and an Fc fragment. Different antibody isotypes have different susceptibilities to digestion (mouse IgG1 being particularly resistant) but adjustments to the practical protocol can take account of these. Since the fragments are monovalent, as compared with the bivalent native antibody, binding affinities will be reduced, which can be a serious problem for antibodies originally of low affinity. When using indirect labeling methods to probe these Fab fragments, care also needs to be taken in the choice of second antibody, anti-Fab label conjugates being necessary rather than anti-isotype label conjugates. However, some apparently specific anti-isotype antibodies can still bind to the CH1 or CL domains of the Fab.

Bivalent Fab fragments ($F(ab')_2$) can be made by treatment of the antibody with pepsin, bromelain or ficin, when the Fc side of the inter-heavy-chain disulfide bond is digested into small fragments. Again, the susceptibility of different isotypes and different species varies considerably, the digestion of rodent antibodies being harder to control than that of rabbit or human antibodies. Mouse IgG2b, for instance, is highly susceptible to complete breakdown by pepsin.

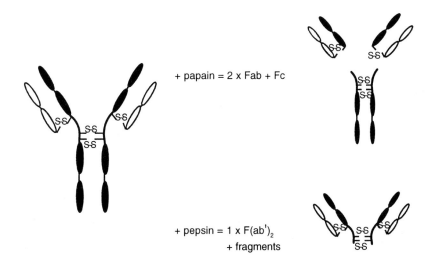

FIGURE 4.2: *Antibody digestion by papain and pepsin.*

4.2.2 Bispecific antibodies

Bispecific antibodies are constructed from more than one source such that they retain the specificities of both original antibodies. They can be made in three ways: by cell fusion methods as described in Chapter 3, by genetic manipulation as will be described in Section 4.3, and by chemical recombination.

The various antibody fragments generated by limited proteolysis in the hinge region are relatively stable and can be rejoined in different combinations without loss of their binding properties. The Fc region is first removed by pepsin digestion leaving a bivalent F(ab')$_2$ fragment. The inter-heavy-chain disulfide bonds are cleaved by a reducing agent such as 2-mercaptoethanol and, after separation by chromatography, the resulting Fab fragments can be joined to different Fab fragments, thus producing a bispecific F(ab')$_2$ fragment. The rejoining can be random, through oxidative reformation of the inter-chain disulfide bonds of the hinge region, or specific, by thioether bonding in which a bis-maleimide linker is used (see *Figure 4.3* and ref. 2). Of particular use in therapeutic applications are constructs of different Fab

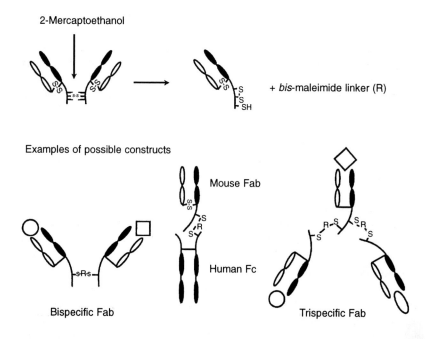

FIGURE 4.3: *Antibody reconstructions by chemical cross-linking.*

specificities where one arm binds to the target epitope, for instance on a tumor cell surface, and the other binds to either a toxin such as saporin (a ribosome-inactivating protein) or an effector cell (T or NK cell) in order to direct cellular toxicity. In both examples, it has been shown that trispecific constructs are even more effective due to tighter binding of effector cell or toxin to the target and enhanced effector activation. The two arms binding to the target need not necessarily recognize the same epitope as long as epitope density is sufficient to allow binding. Careful selection of the tumor target epitope is also important, since simple binding to the target cell is not necessarily enough to internalize the toxin and effect cell killing (see Chapter 7). By similar chemical means, one or more Fc regions of the desired isotype can be reattached to the Fab or Fabs, thus taking advantage of the natural effector mechanisms *in vivo*. This also has the therapeutic bonus of prolonging the survival of the construct *in vivo* (the half-life of normal human IgG is approximately 20 days).

4.2.3 Altering effector functions

For most antibody applications, such as diagnostic assays and antigen purification, the Fc region is unnecessary: its presence or absence has no effect on the binding of antibody to antigen. However, it is important for therapeutic applications, where the body's natural effector mechanisms are to be harnessed. It has been recognized for many years that the various antibody isotypes vary considerably in their ability to harness these effector mechanisms. The use of monoclonal antibody constructs with identical specificities but different Fc regions has aided the accumulation of data on the most important isotypes for particular functions and also as to which residues are involved in these binding mechanisms [3].

The isotypes of mouse, rat and human monoclonal antibodies are not equivalent in their binding characteristics to complement factors and Fc receptors. In analyzing the efficacy of a particular antibody in complement-mediated cell lysis, it has been shown that monoclonal antibodies which work well *in vitro* with heterologous cell targets and complement factors are often poor reagents *in vivo*. Many cell targets have also been found to possess complement restriction factors to inhibit the toxicity of the animal's own complement to its own cells.

The classical pathway of complement activation is initiated by binding of the complement factor C1q to a site on CH2 of human IgG and CH4 of IgM. IgG1 and 3 are the most effective at binding complement; IgG2 is less so and IgG4 is inaccessible to C1q because of steric hindrance caused by the Fab arms. Within the CH2 domain, certain residues have been recognized that are critical for binding. Fc receptors recognize sites in CH2 and CH3 of IgG, although binding to Fc receptors depends on the target cells, which differ in the type and density of Fc receptors (mouse macrophages, for instance, have several different types on the same cell). The results of the effectiveness of various isotypes in binding to these components of the immune response *in vivo* are considered further in Chapter 7.

4.3 Genetic manipulation

The failure of the hybridoma method to reliably produce human monoclonal antibodies suitable for therapy coincided with general technological advances in molecular biology such as DNA sequencing and the polymerase chain reaction (PCR). The definition of antibody gene sequences led the way to genetic manipulation to reshape the original rodent antibody so that it would resemble the human equivalent more closely without losing its specificity. However, the demonstration that antigen-binding Fabs and Fvs could be expressed in bacteria [4,5] led to the generation of entirely human antigen-binding fragments through the construction of combinatorial antibody gene libraries [6,7].

4.3.1 Humanizing rodent antibodies

In the first attempts at genetic manipulation of antibodies, it was thought that the problems associated with human host recognition of foreign protein (see Chapter 7) could be resolved by replacement of the effector part of a specific rodent monoclonal antibody with the equivalent region from a human antibody. The protein domain structure of antibodies is ideal for this purpose, since each domain is coded for by a different genetic exon, such that grafting appropriate exons together will result in a chimeric antibody. In this way, the constant regions could be chosen with the most appropriate effector functions for a particular purpose. This was first done by isolating and sequencing the genes coding for variable (V) regions of the heavy and light chains from a mouse hybridoma and inserting them into separate expression vectors containing the human heavy or light constant (C) region genes derived from myeloma proteins [8]. By co-transfection of these vectors into myeloma or Chinese hamster ovary (CHO) cells, whole antibodies could be expressed producing yields of up to 0.7 g l^{-1} of culture medium. Such chimeric antibodies demonstrated reduced immunogenicity *in vivo* compared with rodent monoclonal antibodies but there still remained some reaction to the rodent part of the molecule.

The next development was to reduce the rodent part of the antibody even further by grafting only the hypervariable regions into the human framework, now known as CDR grafting. Within the antibody variable regions there are regions of hypervariable sequence called complementarity determining regions (CDRs) (see *Figure 4.4*). There are three such regions on each of the heavy and light chains

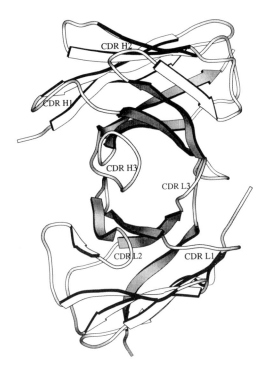

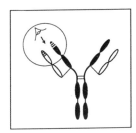

FIGURE 4.4:   *This molecular model shows the six CDR loops of the antigen binding site supported by a β-sheet framework (a single-chain Fv fragment viewed from the front, see inset). Note that the third heavy and light chain (H3 and L3) loops form the center of the binding region. Molecular model (produced using MOLSCRIPT) courtesy of Robyn Malby, Biomolecular Research Institute and CSIRO Division of Biomolecular Engineering, 343 Royal Pde, Parkville, Australia 3052; and reproduced from Hudson, P. (1994) Structure and application of single-chain Fvs as diagnostic and therapeutic agents. In* Monoclonal Antibodies: the Second Generation *(H. Zola, ed.). BIOS Scientific Publishers, Oxford.*

(numbered 1–3). Thus, each arm on the immunoglobulin molecule has six CDR loops (H1, H2, H3, L1, L2, L3) at its tip, supported by a relatively conserved β-sheet sandwich framework. These CDRs are responsible for the binding of the antibody to its specific epitope. Grafting the CDR regions from the specific rodent monoclonal antibody into a human antibody results in the transfer of specific antigen binding, although this is not only dependent on the amino acid sequence of the loops but also on their shape as determined by the framework structure. The framework regions provide the right base for the CDR loops and even single amino-acid residue changes at critical positions can significantly change the loop binding properties through alteration of the structural orientation of the loop. Information obtained from crystallographic analysis and molecular modelling has allowed the design of limited numbers of human frameworks from which the one most similar in structure to the native rodent antibody can be selected. Some loss of the native binding affinities can be expected if CDRs are simply grafted into one of these frameworks, although fine tuning of the loop shape, size and charge composition by mutation of certain residues in the loop or framework can significantly improve binding. More and more information is being accumulated as to which residues are the most important to change or conserve in order to obtain the best binding characteristics. For instance, not all CDRs are equivalent in importance in terms of antigen binding. Indeed, in some identified antibody–antigen reactions, one or two of the CDRs are not directly involved in binding at all. It has been shown repeatedly that the H3 loop is the most diverse of the CDRs, both in sequence and in length (from two to 26 residues), and is always involved in antigen binding, with H2 and L3 being the next in importance.

CDR grafting takes advantage of the many well-characterized rodent monoclonal antibodies that are already available, and the method has undoubtedly produced improved therapeutic reagents which are undergoing extensive clinical trials (see Chapter 7). However, the main thrust of the endeavor to produce new therapeutic products has now shifted to producing entirely human antibodies from combinatorial antibody gene libraries which potentially represent the whole of the antibody gene repertoire as developed *in vivo*.

4.3.2 Combinatorial antibody gene libraries

Combinatorial antibody gene libraries are random combinations of different heavy and light chain genes, inserted into an appropriate expression vector, from which some combinations can be isolated which express protein with antigen binding properties (see *Figure*

4.5). The sources of antibody genes can be hybridomas or B lymphocytes (from spleen, bone marrow or peripheral blood lymphocytes) from immune or even nonimmune individuals from which RNA is isolated, converted by reverse transcription to cDNA and amplified by use of the PCR. Pairs of PCR primers complementary to either end of the antibody genes to be amplified are designed from database antibody sequences, and should amplify a large proportion of the antibody repertoire (see p. 57 and *Figure 4.7*). Following PCR amplification, the heavy and light chain genes of the appropriate size are isolated by preparative agarose gel electrophoresis, purified and their ends subjected to digestion with restriction enzymes. The genes are then cloned into an expression vector (see *Figure 4.5*). Whole antibodies cannot be expressed in bacteria since their assembly requires glycosylation mechanisms that bacteria lack, but it has been demonstrated that antigen-binding antibody fragments (Fab and Fv) can be expressed in *Escherichia coli*

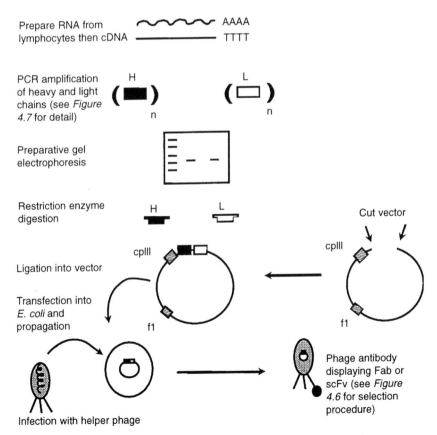

FIGURE 4.5: Combinatorial library construction (see also p. 55).

[4,5], thus enabling the propagation of these genes in prokaryotic cells which are much easier to manipulate than eukaryotic cells. The original combinations of heavy and light chains found in the repertoire of the immune individual from which the RNA was taken are, of course, destroyed by the process of constructing a recombinant antibody gene library. Surprisingly, the random recombination of heavy and light chains does result in antigen-binding molecules, as was first demonstrated by the two leading groups in this technology at the MRC Laboratory of Molecular Biology in Cambridge and the Scripps Research Institute in California [6,7].

Since the initial experiments in constructing combinatorial antibody libraries there have been considerable technological improvements that have enabled the construction of very large libraries (i.e. containing potentially 6.5×10^{10} different recombinants), the isolation of very rare specificities (i.e. $1:10^7$), the production of libraries from nonimmune and semisynthetic sources, and genetic manipulation to improve binding affinities and even change specificity. The most significant technological advance that has promoted these improvements is the packaging of a peptide gene within a filamentous bacteriophage such that the peptide can be expressed on the phage surface [9]. This enables preferential selection of the phage and its genes from millions of others, through the binding of the peptide to its ligand and its subsequent amplification by bacterial infection. Application of the technique to folded proteins and antibody fragments has soon followed.

The phage display technique involves cloning the appropriate genes into a filamentous bacteriophage vector, a single-stranded DNA phage that can infect male *E. coli* (*Figure 4.5*). The phage has three coat proteins (cpIII) displayed at its tip, the function of which is to adsorb to the bacterial sex pilus. If a peptide or protein gene is cloned into the vector as a fusion with the cpIII gene then the peptide or protein is displayed on the surface of the phage. This enables the phage expressing the desired protein (i.e. antibody) to be selected from the mixture of unwanted phage by allowing the specific phage to bind to its ligand (i.e. antigen) and then eluting off the bound phage. This phage which contains the genes coding for the specific antibody is amplified by reinfecting bacteria. If this process, known as panning (*Figure 4.6*), is repeated several times, the original random mixture can be enriched for the specific phage several thousandfold at each round. Finally, individual infected bacterial colonies can be isolated with a very high probability of selecting specific binders. The phage displaying this, now monoclonal, antibody fragment can be used directly in conjunction with an anti-phage labeled antibody, for instance in ELISA (see Sections 3.2.6 and 5.4) or, alternatively, soluble antibody fragments can be generated by simple genetic

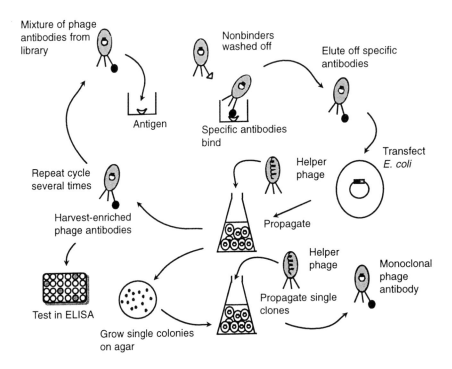

FIGURE 4.6: *Selection from a combinatorial library.*

manipulation according to the methods described below. For diagnostic assays or antigen purification, the constant region of the native antibody is unnecessary as long as the antigen-binding fragment can be detected with a labeled second antibody but, should the antibody be needed for therapeutic purposes which depend on complement or T-cell fixation, a selected constant region can be tagged on from a cloned Fc library and the whole antibody expressed in a eukaryotic cell system.

Methods of library construction. In antibody gene library construction it is important to make as large a library as possible in order to represent as much as possible of the antibody repertoire from which to select the desired clones and increase the chances of obtaining high affinity binders. The major limiting factor on library size is the transformation of the bacteria, which is largely dependent on the competency of the bacterial cells (i.e. the efficiency with which foreign DNA can be inserted into the bacteria, preferably 10^9 or more bacterial colonies per µg of DNA), the method of transfection (preferably electroporation) and the type of vector.

Before the application of phage display, combinatorial libraries were constructed in λ phage vectors by making separate heavy and light chain libraries and then combining the two by appropriate enzyme digestion and religation [6]. The libraries were then screened by filter plaque assays with labeled antigen. This is a standard screening procedure in which DNA from the bacterial colonies on the agar plate is transferred on to nitrocellulose or nylon filters which are then probed with appropriate radiolabeled oligonucleotide probes, thus indicating which colonies contain the corresponding recombinant DNA. The problem with applying this method to antibody libraries was the tedium of screening a minimum of 10^6 recombinants, making it extremely difficult to be able to select for rare specificities. Screening by this method also requires large amounts of antigen (i.e. a library size of 10^7 would need 200 filter lifts), which further restricts universal application. This screening problem was solved with the introduction of phage display vectors.

Some of the first vectors to be used for this purpose were phage vectors (see *Table 4.1*) which have relatively low efficiencies of transformation compared to plasmid vectors such as pUC. Phagemid vectors are plasmids with an origin of replication for filamentous bacteriophage and are 100-fold more efficient than phage vectors. They also contain the cpIII gene and restriction sites to allow for insertion of the foreign DNA. In order to produce free phage particles expressing the protein on the surface, it is necessary to infect the bacteria containing the phagemid DNA with helper phage (i.e. M13KO7). The helper phage has a defective origin of replication so the secretion of the phagemid package is favored. In phage vectors, all copies of the cpIII fusion protein will be displayed on the surface but with phagemid vectors this number can be varied such that all or only one of the coat proteins carries the foreign protein. Normally, a mixture of native cpIII and cpIII-fusion protein expression occurs following helper phage infection, but the ratio can be varied by altering the concentration of isopropylthiogalactoside, which induces expression from the *lac* promoter of the vector. The advantage of

TABLE 4.1: *Common vectors used for antibody phage display*

Vector	Type	Ab fragment	Original ref.
fd-CAT1	Phage	scFv[a]	[7]
fd-tet-DOG1	Phage	scFv	[10]
pHEN1	Phagemid	scFv and Fab	[11]
pComb3 and 3H	Phagemid	Fab	[12]

[a] Single-chain Fv.

monovalent expression is that higher affinity binders will be selected in preference to those of low affinity, whereas multivalent expression would also result in the selection of low affinity binders by virtue of the amplified avidity of binding. For this reason, the fusion of the antibody genes to another coat protein (cpVIII), which expresses multiple copies over the whole phage surface, is no longer favored for antibody selection purposes.

Antibody expression: dAbs, scFvs and Fabs. In 1989 it was demonstrated that VH genes could be isolated from spleen genomic DNA of immunized mice, amplified by PCR, and then cloned into and expressed from *E. coli* as single VH domains [13]. A selection of these were found to bind to the original immunogens (lysozyme and keyhole limpet hemocyanin), despite the loss of the light chain partner, albeit with fairly low binding affinities of up to 5 x 10^7 M^{-1}. These single-domain antibodies, or dAbs as they were designated, were recognized as potential building blocks for higher affinity binders in combination with VL genes. The smaller size of the dAb (14 kDa) potentially enabled it to penetrate tissues more easily and be more accessible to canyon sites of viruses, and thus block activity, and should allow greater resolution in epitope mapping. The disadvantages were the stickiness, due presumably to exposed hydrophobic residues which are normally masked by the light chain, and relatively low binding affinities.

The same researchers went on to develop single-chain Fv (scFv) fragments which are antigen-binding fragments derived from heavy and light chain variable region genes joined together by a short oligonucleotide sequence before cloning into the vector. The protein is expressed as a single chain of the heavy and light chain antigen-binding regions joined together by a flexible linker peptide (see *Figure 4.7* and ref. 7). The vector pHEN1 allows for either the expression of antibody on phage or the soluble expression of scFv fragments through the insertion of an amber stop codon immediately downstream of the antibody insert. Transfection into a suppressor bacterial strain results in phage expression, and the phage antibody can be used in many assay formats when used in conjunction with an anti-phage label. However, in order to take full advantage of the small size of scFv, soluble fragments are preferable, and can be obtained by transfection into a nonsuppressor bacterial strain. The vector also contains a c-myc tag which allows for detection of the scFv by using an anti-c-myc monoclonal antibody label.

Alternatively, entire monovalent Fab fragments can be expressed by amplifying the genes coding for the entire light chain and the VH and CH1 domains of the heavy chain (see *Figure 4.7*) [12]. The 3'

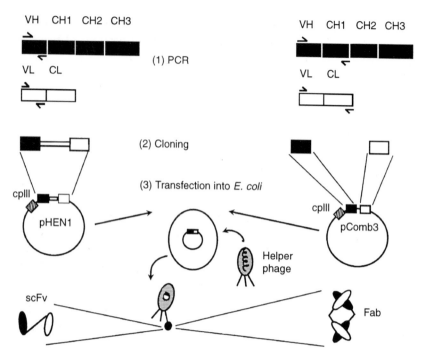

FIGURE 4.7: *(1) The antibody genes coding for the VH and VL regions (scFv) and VH-CH1 and VL-CL regions (Fab) are amplified by PCR (using specific antibody primer pairs complementary to the areas shown by arrows). (2) Cloning: The heavy and light chain genes are purified and digested by restriction enzymes compatible with cut sites in the appropriate vector. For scFv construction, the antibody genes are first joined by a (Gly$_4$Ser)$_3$ linker. For Fab construction, the genes are cloned separately and sequentially into the vector. (3) E. coli are transfected with the cloned antibody genes by electroporation, and propagated. Infection with helper phage allows for the packaging and expression of phage carrying recombined antibody genes with Fab or scFv expressed on the surface.*

PCR primer of the heavy chain (one for each isotype) hybridizes to a hinge region sequence which includes a Cys involved in the heavy–light chain disulfide bond. This bond later assists in the association of the two chains in the bacterial periplasm, so there is no need to construct an artificial linker as with the scFv. It is customary to clone first the light chain and then the heavy chain genes into the vector so that only the heavy chain is linked to the cpIII gene. Once selection of the desired phage antibody has been made, soluble Fab fragments can easily be made by cutting out the cpIII gene with restriction enzymes and religating, although this does result in reduced yield (1–2 mg l^{-1} of culture medium). The Fab fragment is said to be more stable than scFv and better resembles the natural antibody in its flexibility and therefore in assisting in the induced fit of antibody–antigen interaction. Detection of Fabs is simplified in that commonly available, inexpensive anti-Fab label conjugates can be employed.

The larger the combinatorial library the more diverse will be the range of antibody specificities and affinities obtainable. As mentioned earlier, the size of the library is dependent on the transfection efficiency of the bacterial strain which is currently limited to 10^9. A new method of increasing this size to 6.5×10^{10} and potentially 10^{13} has been demonstrated, known as combinatorial infection [14]. Very diverse heavy chain libraries were made in a plasmid by pooling heavy chain variable region gene banks combined with random heavy chain CDR3 sequences. Similarly, a light chain library was made in a phage and allowed to infect bacteria already containing the heavy chain plasmid. Each vector contained *lox* P sites which allow recombination of the two vectors when the bacteria are co-infected with a phage coding for the enzyme Cre recombinase. The result is that the heavy and light chains are combined in the same replicon and the library size is potentially equivalent to the number of bacteria infected.

Naive and synthetic gene libraries. Naive libraries are those derived from a nonimmune source. The advantage of these is that immunization can be bypassed and the repertoire, if large enough, can be screened for binding of virtually any antigen, including self-antigens, which is very difficult to achieve by conventional methods. It has also been suggested that antibodies derived from naive libraries are likely to have a different spectrum of specificities than those hitherto obtained from immunized sources where immuno-dominant epitopes are favored at the expense of the less immuno-genic. By panning phage libraries against different antigens, wide varieties of antibody specificities have been isolated from single naive libraries with binding affinities comparable to those expected from a primary immune response (10^6–10^7 M^{-1}).

Specific antibodies have also been obtained from semisynthetic libraries. These have originated from pooled gene libraries or single clones which have been subjected to recombination with randomized nucleotide sequences in the CDR H3 region followed by selection against alternative antigens. This strategy has resulted in the conversion of an anti-tetanus toxoid antibody to one specific to fluorescein [15], which demonstrates that modulation of H3 sequences can be sufficient to transform the specificity of an antibody binding site. Similarly, a very large repertoire of VH genes has been assembled, as scFv, from pooling a bank of 50 cloned human VH gene segments. This, combined with random nucleotide sequences encoding CDR H3, produced a 'single-pot' library from which antibody specificities were isolated against (initially) 18 different, diverse antigens including haptens, self- and foreign antigens and intracellular proteins [16]. Similar library diversity has also been demonstrated by others.

Refining antibody characteristics. One important advantage of the combinatorial library approach over the hybridoma method is the ease and speed with which the initial antibody can be changed and improved upon without having to begin again. Once a library has been established it can be screened again, within days, for other antibodies with different binding specificities. Cross-reactivity of antibodies with other antigens can be reduced at the panning stage by introducing the alternative antigen in the soluble phase or negative panning against cross-reacting antigens. Similar strategies can also be employed to increase cross-reactivity which might be desirable under circumstances that require the recognition of rapidly evolving viral antigens.

Typical antibody–antigen binding affinities of recombinant antibodies from immune sources fall within the range of 10^7–10^9 M^{-1}, whereas from nonimmune sources the range tends to be 100-fold lower. Having the antibody genes conveniently packaged, however, enables the *in vitro* manipulation of these affinities (and also specificities) by a number of different strategies. The simplest of these is probably chain shuffling in which one of the cloned chains is retained and recombined with the original library of the complementary chain. This can produce a family of specific antigen binders, some of which may have improved binding characteristics. Indeed, up to 300-fold improvement of binding affinities over the original antibody have been reported. It has been shown that the heavy chain is the most promiscuous in that a given heavy chain can recombine with light chains of an unrelated specificity and still retain binding to its original antigen, whereas the converse has not been shown for light chains.

Random or site-directed mutagenesis of the variable region genes is another strategy that can be effected by standard chemical means, by PCR or by *in vivo* mutagenesis using mutator bacterial strains. As more antibody combining sites are being analyzed, knowledge is accumulating as to the most appropriate sites to conserve or change in this respect (see Section 4.3.1).

Natural antibodies are bivalent whereas the recombinant antibody fragments so far described are monovalent, and therefore have an inherent disadvantage in terms of binding avidity. To overcome this problem, bivalent miniantibodies based on scFv fragments have been constructed in *E. coli* [17]. Two scFvs are dimerized through attachment to each of the flexible hinge regions which are joined by a four-helix bundle motif. The dimer is similar in size to a monovalent Fab fragment. This type of construct also facilitates the construction of bispecific antibodies by combining scFvs of different specificities. In this way, one could attach an effector function by combining a scFv specific for a T-cell epitope or complement factor.

Other useful antibody constructs that take advantage of phage display technology are Fabs fused to cpIII and expressed on the phage surface, which also express alkaline phosphatase fused to the more abundant coat protein cpVIII. Thus, a complete ELISA reagent is produced by the bacteria. Clearly, other such constructs and fusion proteins are possible, if the gene sequence coding for the protein tag is known and can be expressed in bacteria without being toxic to the bacterial growth or requiring special processing that the bacteria cannot provide.

Other antibody expression systems. Bacterial expression systems are very convenient for the production of antigen-binding fragments because of the ease and speed with which protein products can be made. However, whole antibodies cannot be assembled in bacteria because of their inability to perform the N-glycosylation of the CH2 domains. Mammalian cells, including plasmacytoma, myeloma and CHO cells, have been used for this purpose. Antibody gene fragments selected from bacterial systems need to be recloned into appropriate vectors containing the processing codes necessary for eukaryotic expression, but whole antibody yields of the order of several μg per ml, and even 500 μg ml^{-1}, have been reported.

Human antibodies for therapeutic applications have been expressed in transgenic mice. Cloned immunoglobulin V, D, J and C genes can be introduced into the mouse genome through microinjection of pronuclei in a fertilized mouse egg. The offspring showing the presence of the transgene can give rise to mouse colonies producing human antibody genes. Immunization of these mice will produce polyclonal and monoclonal antibodies recognized as human. Doubt has been expressed, however, as to the size of antibody repertoires that can be obtained by this method due to limits in the amount of foreign DNA that can be carried by the mice.

Recombinant antibodies can also be expressed in whole plants, and in protoplast or callus or liquid suspension. Although regeneration of antibody-producing plants can take several months, the processes are relatively simple, and expression and evaluation of recombinants from protoplast preparations can be done within days. In addition, plant propagation might provide a cheaper, large-scale production facility for therapeutic monoclonal antibodies, since antibody levels can reach 1% of total extractable protein from regenerated plants, as long as the immunogenicity of plant-derived products does not prove problematic. Indeed, a whole new application for antibodies can be predicted in the *in vivo* control and analysis of plant metabolism and development.

4.4 Summary and future prospects

This chapter has described the wide variety of antibody fragments and derivatives that can be obtained by chemical and genetic manipulations. The last 5–10 years have seen great advances in the technology of antibody genetic engineering, which is still in an exciting and rapid phase of development. There is currently considerable debate as to the relative merits of making Fab or scFv fragments. The technology is still in its infancy and, no doubt, as more antibodies are made by both methods and the techniques are further refined, a preferred method for a particular application or family of antigens will emerge based on convenience and reliability. Individual antibodies made by either method are proving to be useful and possibly unique additions to the current repertoire of diagnostic and therapeutic reagents (see Chapter 7). Whatever the initial outcome in terms of specificity or binding affinity, the isolation of antibody genes enables the fine tuning of desired characteristics which cannot be done by the conventional hybridoma method. In addition, the ability to model the antibody binding site from gene sequences and the analysis of how changes in those sequences can affect binding both help to define the structure of the epitope on the antigen.

Although the major impetus for genetic engineering of antibodies is undoubtedly the need to develop good human therapeutic reagents where more conventional methods have failed, this technology should not be overlooked for other applications. For instance, the genes of hybridoma antibodies already developed can be rescued and improved upon; immunization is not always possible in rodents, if the immunogen is toxic; homologous antibodies may be required in other species for therapy (antibody genes can be manipulated for any species if suitable antibody primers can be designed); in some applications, a great diversity of antibody specificities is desired which would be difficult and time consuming to screen for by hybridoma methods. The speed and flexibility of the genetic approach has overwhelming attractions. Although progress in the technology is firmly in the hands of a few expert laboratories, the production of useful antibodies by phage display can now be achieved by others with basic molecular biology facilities. One can envisage, in the not too distant future, that a desired antibody will be routinely designed by computer and selected from a family of antibody gene libraries obtained from the shelf.

References

1. Roitt, I., Brostoff, J. and Male, D. (1993) *Immunology* (3rd edn). Gower Medical Publishing, London.
2. Glennie, M.J., McBride, H.M., Worth, A.T. *et al.* (1987) *J. Immunol.,* **139**, 2367.
3. Tao, M.H., Smith, R.I.F. and Morrison, S.L. (1993) *J. Exp. Med.,* **178**, 661.
4. Skerra, A. and Plückthun, A. (1988) *Science,* **240**, 1038.
5. Better, M., Chang, C.P., Robinson, R.R. and Horwitz, A.H. (1988) *Science,* **240**, 1041.
6. Huse, W.D., Sastry, L., Iverson, S.A. *et al.* (1989) *Science,* **246**, 1275.
7. McCafferty, J., Griffiths, A.D., Winter, G. and Chiswell, D.J. (1990) *Nature,* **348**, 552.
8. Riechmann, L., Clark, M., Waldmann, H. and Winter, G. (1988) *Nature,* **332**, 323.
9. Smith, G.P. (1985) *Science,* **228**, 1315.
10. Clackson, T., Hoogenboom, H.R., Griffiths, A.D. and Winter, G. (1991) *Nature,* **352**, 624.
11. Hoogenboom, H.R., Griffiths, A.D., Johnson, K.S., Chiswell, D.J., Hudson, P. and Winter, G. (1991) *Nucl. Acids Res.,* **19**, 4133.
12. Barbas, C.F. III, Kang, A.S., Lerner, R.A. and Benkovic, S.J. (1991) *Proc. Natl Acad. Sci. USA,* **88**, 7978.
13. Ward, E.S., Gussow, D., Griffiths, A.D., Jones, P.T. and Winter, G. (1989) *Nature,* **341**, 544.
14. Griffiths, A.D., Williams, S.C., Hartley, O. *et al.* (1994) *EMBO J.,* **13**, 3245.
15. Barbas, C.F. III, Bain, J.D., Hoekstra, D.M. and Lerner, R.A. (1992) *Proc. Natl Acad. Sci. USA,* **89**, 4457.
16. Nissim, A., Hoogenboom, H.R., Tomlinson, I.M. *et al.* (1994) *EMBO J.,* **13**, 692.
17. Pack, P., Kujau, M., Schroeckh, V. *et al.* (1993) *Bio/Technology,* **11**, 1271.

5 Diagnostic Applications of Antibodies

5.1 Immunoprecipitation reactions

The reaction between an antibody and its antigen can be directly observed *in vitro* at the point of equivalence of the concentrations of these components. At this point an immunoprecipitate is observed.

The property of immunoprecipitation has been widely exploited for many years for the detection and quantitation of antigens and antibodies [1]. Such immunochemical reactions have been most commonly carried out in semi-solid matrices such as agar gels. The diffusion of the immunochemical components through such gels gives rise to a concentration gradient which ensures that, at some point, conditions for immunoprecipitation will be achieved provided that immune complex lattices are formed. The position and nature of such immunoprecipitates in the gel can be used as a measure of the concentrations of the immunocomponents.

A qualitative indication of immune complex reaction is seen in double immunodiffusion. In its simplest form, a thin slab of agar gel is prepared with two small wells into which small volumes of the antibody and antigen solutions are placed. Diffusion is allowed to take place typically for 24 h. Immune complex formation is indicated by the presence of a white precipitin line formed at some point between the two wells (*Figure 5.1a*). A more complex version of this is represented by the Ouchterlony plate in which wells can be formed in a circular format with a further well in the center. Various dilutions of a given antiserum can be placed in the outer wells and a standard antigen solution placed in the central well. Assessment of antibody titer can then be made corresponding to the dilution at which a precipitin band is seen (*Figure 5.1b*).

(a)

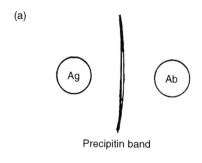

Precipitin band

(b)

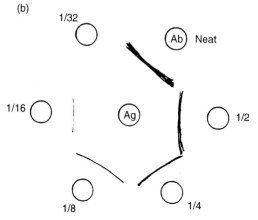

FIGURE 5.1: *(a) Double immunodiffusion involves placing a few microliters of antigen (Ag) and antibody (Ab) in adjacent wells formed in a thin slab of gel (e.g. agar). If the antibody and antigen cross-react, a band of immunoprecipitate forms at the zone of equivalence. (b) Ouchterlony immunodiffusion consists of a well containing antigen surrounded by wells equidistant from the center well into which various dilutions of antiserum are placed. This gives some guide as to the 'titer' of the antiserum.*

A more quantitative form of these techniques is single immunodiffusion. Here, only one of the immunocomponents is allowed to diffuse, the other being homogeneously incorporated into the gel. In this situation, antigen in a well is allowed to diffuse through antibody-containing agar so as to form a ring of precipitation. The area within the precipitin ring is proportional to the antigen concentration (*Figure 5.2a*). Thus, unknown antigen concentrations can be determined by comparison with standard antigen solutions set up on the same plate. Antibody titers can be determined in a similar manner if antigen is incorporated into the gel and antiserum placed in the wells. A more commonly used form of this assay is not based on passive diffusion but rather on electrophoresis. Here, an element of quantitation is introduced since the distance moved by the precipitin band is proportional to the concentration of immunocomponent in the well. For obvious reasons the technique is known as rocket immuno-electrophoresis (*Figure 5.2b*). Immunoelectrophoretic methods are generally more rapid and sensitive than passive diffusion techniques. All of these techniques benefit from the use of protein staining methods to highlight the presence of precipitin bands, for example by use of Coomassie blue dye.

Many types of immunoprecipitation assay have been described but the examples above are representative of the general principles of all of them. Since the formation of immunoprecipitates is observed directly in all of these methods, their sensitivity is severely limited and in the best cases is in the range 20 µg ml^{-1} to 2 mg ml^{-1}. Greater

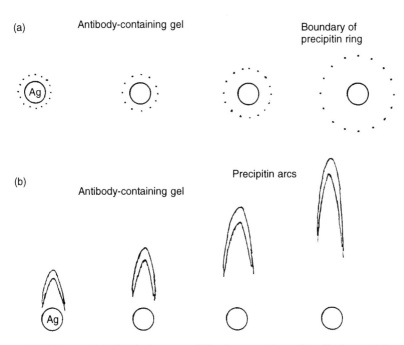

FIGURE 5.2: (a) Single immunodiffusion consists of wells formed in a gel slab into which antibody has previously been incorporated as part of the gel whilst liquid. When antigen (Ag) solutions are placed in the wells the area encompassed by the precipitin ring is proportional to the amount of antigen. (b) Rocket immunoelectrophoresis is set up in the same way as (a) except that the proteins migrate under the influence of an applied electric field rather than by diffusion. This permits greater sensitivity to be achieved. Both these systems can be reversed by incorporating a given amount of antigen in the gel in order to characterize antibody in the wells.

sensitivity can be obtained in solution-phase immunoprecipitation reactions where smaller amounts of precipitin can be detected by turbidimetry or nephelometry, and particularly where immune complex formation can be detected by particle agglutination.

Early agglutination assays made extensive use of erythrocytes which were coated with antigen. Later methods used latex microparticles because of their greater robustness. Such reagents can be used as a means of detecting antibody (agglutination) or soluble antigen (agglutination inhibition) [2]. In the former, the presence of antibody causes antigen-coated particles to cross-link and sediment as a diffuse layer whereas, if antibody is absent, the particles sediment as a tight pellet at the bottom of the test-tube. For the detection of soluble antigen, a known amount of antibody sufficient to cause agglutination is also introduced as a reagent in the knowledge that introduction of a soluble antigen will result in the formation of solution-phase immune complexes and thus inhibit agglutination. Other architectures can also be envisaged (*Figure 5.3*). Assays of this type are often capable of yielding sensitivities below 1 μg ml^{-1}.

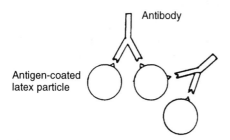

Antibody

Antigen-coated
latex particle

FIGURE 5.3: *Schematic diagram of latex agglutination in which cross-reacting antibodies can be detected by their ability to cross-link latex particles coated with the relevant antigen. Such 'agglutination' can be detected visually.*

5.2 Radioimmunoassay

The poor sensitivity of detection of the immune complexes is the principle limiting factor in the use of the immunoprecipitation techniques described in Section 5.1. Greater sensitivity can be attained by the use of radioimmunoassay (RIA) techniques. In its original form, the RIA described by Yalow and Berson [3] involved the determination of insulin concentration in a serum sample. The serum sample was incubated with a limiting amount of anti-insulin antibody

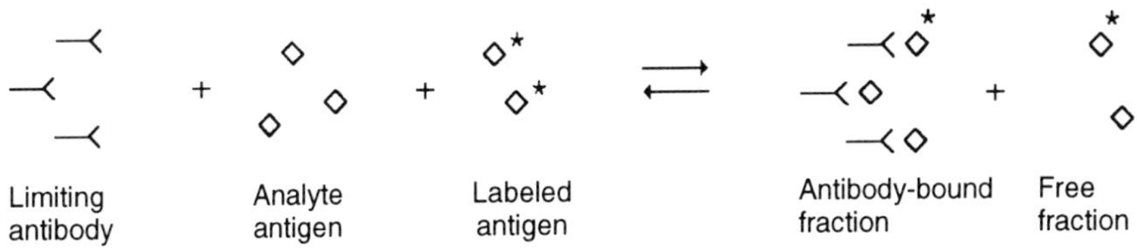

| Limiting antibody | Analyte antigen | Labeled antigen | Antibody-bound fraction | Free fraction |

FIGURE 5.4: *Schematic representation of RIA.*

together with a trace amount of insulin which had previously been labeled with radioactive iodine (^{131}I) (*Figure 5.4*). Thus, because the labeled antigen and sample antigen were forced to compete for the limited number of antibody binding sites, the amount of radioactivity bound in the form of immune complexes was inversely proportional to the amount of antigen (insulin) present in the sample. Naturally, it was necessary to separate antibody-bound and free antigen prior to measuring immune complex radioactivity in a γ counter. In this case separation was achieved by paper chromato-electrophoresis.

This basic methodology has been refined over many years and used widely. The most commonly used radioactive labels have been iodine-125 (^{125}I) and tritium (^{3}H). The former is a high energy γ emitter with a high sensitivity of detection and short half-life (60 days), whereas the latter is a low energy β emitter with a low sensitivity of detection and a long half-life (12 years).

5.2.1 Radioactive labeling

^{125}I has been used widely for protein labeling since it can be incorporated into tyrosine and histidine residues [4]. Since it is not part of the normal protein structure it is known as an external label in this situation. ^{3}H has most often been used for the labeling of small molecules such as steroids where it can be used in place of normal hydrogen atoms; in such cases ^{3}H is described as an internal label. Since the sensitivity of detection of ^{125}I is so superior to ^{3}H, attempts are often made to use ^{125}I as a label for small molecules. This works quite well in the case of the thyroid hormones since they naturally contain iodine which can be replaced with ^{125}I as an internal label. However, this is a special case and ^{125}I can generally only be used as an external label. Often, the small molecule concerned must be modified in order to accept an atom of ^{125}I. This has been done by chemical introduction of tyrosyl- and histidyl-type side-chains into the molecule to be labeled. This represents a major chemical change to the molecule and often results in problems with subsequent antibody recognition. However, this approach has been successfully used in several cases [5].

Whilst introduction of ^{3}H requires dedicated synthetic procedures, the introduction of ^{125}I is relatively straightforward if the molecule to be labeled possesses the appropriate acceptor moities. Several variations of a basic procedure exist. All the methods rely on the use of the radioactive iodide anion (^{125}I$^-$) which is usually in the form of a sodium iodide solution. The ^{125}I$^-$ is oxidized prior to the addition of the molecule to be labeled or in the presence of the molecule to be labeled. The molecular iodine formed then reacts spontaneously and rapidly with tyrosyl or histidyl side-chains to yield the labeled substance. The reaction is then stopped most often by the introduction of a reducing agent to re-form unreactive ^{125}I$^-$. The labeled product is separated from the mixture by an appropriate technique. If the labeled molecule is a protein, it is most conveniently separated from the lower molecular weight components by using a small gel filtration column. If the labeled molecule itself is relatively small, it may be possible to purify by thin layer chromatography or high-pressure liquid chromatography.

The most widely used methods of iodination use quite different oxidants. The earlier method involves mixing solutions of ^{125}I$^-$ (1 mCi) and the substance to be labeled with a solution (10 μl) of chloramine T (2.5 mg ml^{-1} in phosphate buffer, 0.2 mol l^{-1}, pH 7.4). After approximately 10 sec the reaction is stopped by addition of a solution (10 μl) of sodium metabisulfite (6 mg ml^{-1} in phosphate buffer, 0.2 mol l^{-1}, pH 7.4) prior to purification. The total reaction volume is usually less than 100 μl. The most recent and widely used method involves the use of a solid-phase oxidant. Here, a small glass/plastic bead coated with the oxidant (for example, iodogen) is added to a solution of the ^{125}I$^-$ and a solution of the substance to be labeled is introduced. After a few minutes, the mixture is removed from the bead and purified to yield the labeled product. This latter method is somewhat milder than the former method and is particularly favored in situations where the substance to be labeled is susceptible to damage by oxidizing or reducing agents.

^{125}I may also be introduced into proteins and peptides by indirect methods. The most widely used of these methods was originally described by Bolton and Hunter [6]. Here, the N-hydroxysuccinimide ester of 3-(4-hydroxyphenyl)propanoic acid is firstly labeled with ^{125}I by an oxidative method and purified. The appropriate amount of this compound is then mixed with a solution of the substance to be labeled whereupon it spontaneously couples to primary or secondary aliphatic amines, such as lysine side-chains. In this way, the substance is never exposed to oxidizing conditions which may otherwise be deleterious. Furthermore, the method has a different specificity, e.g. lysine residues or N-terminal amines as opposed to tyrosine residues, which may be advantageous under certain circumstances.

5.2.2 Separation systems

An important area in the technique of immunoassay has been the development of efficient methods for separating immune complexes from the reaction mixture prior to quantitation of bound activity. Clearly, a more convenient technique than electrophoresis is required for routine application. One of the first separation methods used involved addition of a small volume of charcoal suspension to the reaction mixture. Sedimentation of the charcoal then allowed supernatant to be removed and the pellet isolated. Free small molecules bind avidly to charcoal whereas, when the small molecule is bound to an antibody, it does not bind to the charcoal. In this way, the activity of labeled reagent as measured in the charcoal pellet represents unbound material whereas activity in the supernatant represents activity bound in the form of immune complexes. Despite the relative simplicity of this method it has several drawbacks, not least being the ability of charcoal to bind only relatively small molecules such as haptens and small peptides.

An alternative method to the use of charcoal involves preferential precipitation of the immune complexes followed by separation and quantitation. One method of doing this is to make use of immunoprecipitation by introducing a second antibody having the appropriate specificity for the primary reagent antibody. In this way, secondary immune complexes are formed which, on precipitation, coprecipitate the antigen. Given the requirements for immuno-precipitation, the concentration of second antibody used must be carefully determined and it is also necessary to have present an appropriate amount of carrier immunoglobulin from the same species as the primary antibody. Such precipitations can be enhanced and made more robust by the introduction of a solution of polyethylene glycol (PEG) (~1%, w/v). The need to use a second antibody to bring about precipitation can be obviated by precipitating large molecular weight immune complexes by a more concentrated solution of PEG (~13%, w/v).

The above methods are somewhat inconvenient and have gradually been replaced by the use of solid-phase reagents in which, for example, antibody is immobilized on to microparticles or macro solid-phase matrices such as coated beads, tubes or microwells (see Section

2.4.2). Generally, coupling to microparticle suspensions such as cellulose or amino-silane-coated magnetizable particles has been performed using chemical methods, whereas coupling to surfaces has relied on passive adsorption. *Table 2.2* summarizes the advantages and disadvantages of these methods.

5.3 Immunoradiometric assays

In 1968, Miles and Hales [7] described an alternative to RIA and introduced the technique of immunoradiometric assay (IRMA). This technique differs from RIA in several respects. Firstly, the antibody itself is labeled as distinct from the labeled antigen of RIA and, secondly, the IRMA is a noncompetitive reaction in contrast to RIA. Moreover, excess antibody, rather than limiting antibody, is used.

In its original form the immunoradiometric assay involved incubation of the sample antigen with excess ^{125}I-labeled antibody. In this first incubation, all of the available sample antigen was converted to labeled immune complexes. Excess labeled antibody was 'mopped up' by incubation of the reaction mixture with a solid-phase antigen derivative; in this case, using cellulose as the solid support (*Figure 5.5*). Separation of the solid-phase–labeled antibody complexes from the solution-phase immune complexes thus permitted the subsequent quantitation of radioactivity in either of these phases. The activity measured in the solid phase was thus inversely proportional to the amount of analyte antigen originally present in the sample. This form

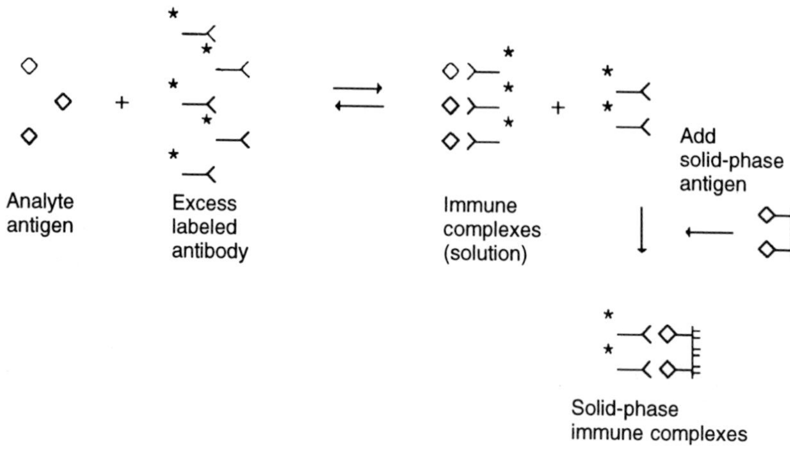

FIGURE 5.5:   *Schematic representation of classical IRMA.*

of IRMA was not used widely despite its proposed theoretical and practical advantages. Part of the reason for this was the requirement for large quantities of highly purified labeled antibodies, in contrast to RIA, which generally required the use of only very small amounts of crude antiserum. Furthermore, large quantities of solid-phase antigen were required which presented practical difficulties in situations where supplies of appropriate antigen for solid-phase derivatization were scarce or expensive.

A variation of the IRMA was subsequently introduced and became known as the 'two-site' immunoradiometric assay [8]. This technique relies on the ability of large antigens to bind simultaneously more than one molecule of antibody. This occurs when such antigens possess several distinct epitopes or repeating epitopes. In its most basic form, this assay uses two reagents; that is, labeled antibody and solid-phase antibody. It is obvious that the specificities of these two antibody populations should not conflict with each other and compete for the same epitopes. Thus, incubation of analytical sample with solid-phase and labeled antibody results in the formation of solid-phase–antigen–labeled antibody 'sandwich' immune complexes if antigen is present in the sample (*Figure 5.6*). When the solid phase is separated and the activity quantified, the latter is directly proportional to the amount of antigen present in the original sample.

Whichever form of IRMA is used, it is apparent that high levels of performance require the use of highly purified labeled antibodies. Such reagents can only be obtained from polyclonal antibody populations by immunoaffinity purification (see Section 8.1).

The advent of monoclonal antibody technology [9] resulted in much greater use of the two-site assay since it became possible to obtain large quantities of chemically pure, well-characterized antibodies. Presently, such assays are often the methods of choice for the

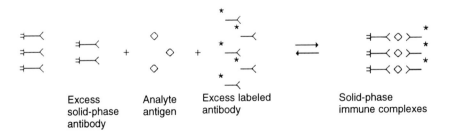

Excess solid-phase antibody   Analyte antigen   Excess labeled antibody   Solid-phase immune complexes

FIGURE 5.6: *Schematic representation of a two-site IRMA.*

quantitation of large antigens. These assays tend to be much more rapid, sensitive and robust than conventional competitive binding immunoassays as exemplified by RIA. The limitation of two-site assays is the size of the antigen; it is not possible to configure such assays for small molecules. For peptides, it becomes increasingly difficult to select appropriate antibodies and configure two-site assays for molecules having molecular weights of below a few thousand daltons.

5.4 Other immunoassay architectures

RIA is an example of a competitive binding assay which uses labeled antigen and limiting antibody as the binding reagent. It is also possible to set up a competitive binding assay using limiting labeled antibody and solid-phase antigen as a competing species with the analyte antigen (*Figure 5.7*). Such a method may be desirable when a pure antibody is available for labeling. This can often be a simpler process than labeling and purifying a small antigen or analog as is required in the classical immunoassay architecture.

Another type of labeled antibody assay is used when the analyte itself is an antibody. Such assays are used widely for allergy testing and for infectious disease testing where ingress of a foreign substance or organism has produced an allergic response (IgE) or an immune response (IgM or IgG), respectively. Here, a solid-phase antigen is used to capture antibody if it is present in the analytical sample. The presence of bound antibody is then established by a second incubation using an anti-Ig-labeled antibody. This architecture has formed the basis of commonly used RAST and ELISA tests (*Figure 5.8*).

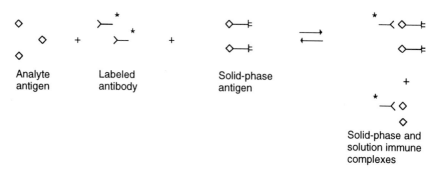

Analyte antigen

Labeled antibody

Solid-phase antigen

Solid-phase and solution immune complexes

FIGURE 5.7: *Schematic representation of a competitive binding immunometric assay.*

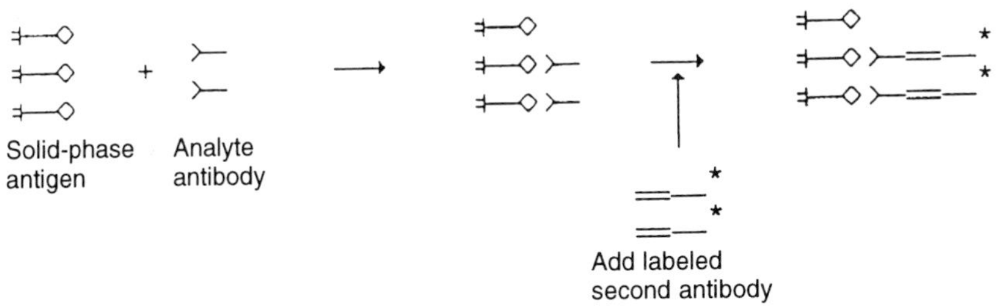

FIGURE 5.8: *ELISA architecture where the analyte itself is an antibody.*

A plethora of architectures have been described where the primary antibody is neither labeled nor used as a solid-phase reagent. Here, use is made of indirect methods where, for example, an anti-species Ig antibody is used either as a capture or a labeled reagent (*Figure 5.9*). A variation of this theme is to derivatize one of the immuno-components in such a way that it is bound by a corresponding binding agent. The most widely used example of this technique is to derivatize the primary antibody with biotin and then use streptavidin either in a solid-phase or labeled form as a 'universal' reagent (*Figure 5.10*). This method relies on the fact that streptavidin binds biotin with an extremely high affinity.

5.5 Nonisotopic immunoassays

Early immunoassays and immunometric assays utilized radioisotope labels, principally in the form of ^{125}I or ^{3}H. Disadvantages of these labels include the short half-life (60 days) of the former and low specific activity of the latter. These factors, coupled with the emotive bias against the use of radioisotopes, stimulated the search for

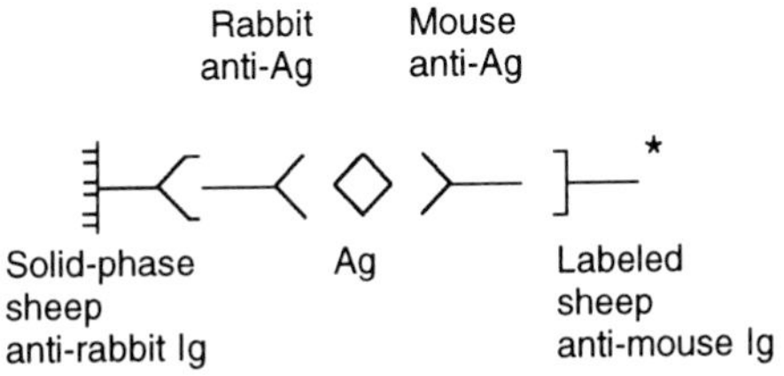

FIGURE 5.9: *Use of indirect labeled and solid-phase antibodies as 'universal reagents'.*

Biotinylated antibody immune complexes

Solid-phase avidin

Solid-phase complexes

FIGURE 5.10: *Use of the biotin–avidin system for the separation of immune complexes.*

nonradioactive alternatives. Such alternatives are gradually becoming more popular though radioisotopes are still used widely.

5.5.1 Enzyme immunoassay

The technique of labeling antigens and antibodies with enzymes [10] formed the basis of one of the earliest forms of nonisotopic immunoassays. The immunochemical reaction is performed in the same way as for a RIA or IRMA. However, due to the relatively unstable nature of enzymes, it is necessary to use a method of separating bound and free fractions which does not prevent the enzyme from yielding the end-point. Thus, solid-phase systems such as antibody- or antigen-coated tubes or beads have found particular application in enzyme immunoassay since this form of separation does not result in protein denaturation.

Quantitation of the bound or free phase is performed by measurement of the associated enzyme activity. Classically, this is done by adding a colorless substrate system which when acted on by the enzyme yields a colored product which can be quantified spectrophotometrically. Thus, when separation of bound and free phases has been achieved, substrate solution is added and, after a predetermined time, the reaction is stopped to permit measurement of the optical density of the product. This gives a measure of the amount of antigen present in the original sample. *Table 5.1* gives examples of a number of enzyme systems that have been used.

Conventional enzyme reactions are often incapable of yielding sufficiently high sensitivities of detection for many immunoassays and so various attempts have been made to improve the detectability of such systems. One very elegant method involves the use of enzyme amplification, in which the product of the primary enzyme reaction is used to set up a 'futile cycle' in a second enzyme system capable of generating enormous quantities of product [11]. An example of this is

TABLE 5.1: *Examples of some enzymes and substrates used in enzyme immunoassay*

Enzyme	Source	Substrate	End-point
Alkaline phosphatase	E. coli and veal intestine	p-Nitrophenyl phosphate	Chromogenic
		NADP/secondary enzymes	Amplified
		AMPPD	Chemiluminescent
Peroxidase	Horseradish	OPD	Chromogenic
		ABTS	Chromogenic
		TMB	Chromogenic
		Luminol	Chemiluminescent
β-Galacto-sidase	E. coli	p-Nitrophenyl-β-galactose	Chromogenic
		MUG	Fluorogenic

AMPPD, adamantyl-1,2-dioxetane phenyl phosphate; OPD, o-phenylene diamine; ABTS, 2,2-azino-di(3-ethylbenzothiazoline-6-sulfonate); TMB, 3,3',5,5'-tetramethylbenzidine hydrochloride; MUG, 4-methylumbelliferyl-β-D-galactose.

shown in *Figure 5.11*. The high levels of sensitivity attainable with this system are, however, only gained by a significant increase in chemical complexity.

Measurement of optical density is not an ideal means of quantifying small quantites of enzyme because of the limitations of sensitivity and dynamic range inherent in such measurement. Improvements in

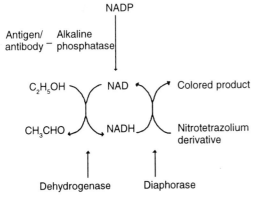

FIGURE 5.11: *Example of an enzyme 'futile cycle' for use in amplified enzyme immunoassay. The enzyme label alkaline phosphatase hydrolyzes NADP⁺ to NAD⁺ which then takes part in a 'futile cycle' to generate enormous quantities of a colored product.*

enzyme quantitation have been made by using the enzyme to modulate luminescent processes such as fluorescence, chemiluminescence and bioluminescence. The first of these is most readily appreciated in that a nonfluorescent substrate is chemically modified to yield a fluorescent substrate which can be quantified in a spectrophotofluorimeter. Such methods are inherently more sensitive and have greater dynamic range than optical density measurements.

The use of enzymes to modulate chemiluminescent and bioluminescent reactions offers a considerable number of possibilities. One of the first examples of this type of technique involved the use of horseradish peroxidase-labeled antigens and antibodies to catalyze the chemiluminescent reaction of hydrogen peroxide and luminol. In an analogous way, one of the first chemiluminescent immunoassays used the ability of hematin to catalyze the luminol reaction [12]. In these systems, the bound or free immunochemical fraction is treated with solutions of hydrogen peroxide and luminol, and the intensity of chemiluminescence emission quantified using a high efficiency photomultiplier (luminometer). This form of luminescence can generally be more sensitively quantified than simple fluorescence. However, this particular chemiluminescent system suffers from a relatively poor quantum yield and was not used widely until it was found that the yield could be dramatically improved by the introduction of 'enhancers' such as p-iodophenol [13]. The exact mechanism of these reactions is, however, still not fully understood.

A more rational approach has been to synthesize nonchemiluminescent precursors which become chemiluminescent upon exposure to a relevant enzyme. An example of this is the use of the enzyme alkaline phosphatase to cleave the phosphate group from a protected, stabilized dioxetane in order to convert it to a dephosphorylated compound which spontaneously emits light [14]. A similar technique has been used with phosphorylated firefly luciferin which will only undergo a light emitting reaction following dephosphorylation.

5.5.2 Fluorescence immunoassay

The use of fluorescent compounds to label antibodies as immunofluorescent microscopy reagents has been popular for many years. Fluorescent labels have also been used as the basis of fluorescence immunoassay. Whilst fluorescent end-points have been discussed above in the context of enzyme immunoassay; here, we are concerned with the use of the fluorescent compound itself to label the antigen or antibody. In this way, it is possible to quantify immune complex formation on the basis of the intensity of fluorescence of the label present in the bound or free fraction of the immunochemical reaction mixture. Such measurements are readily made in spectrophotofluorimeters.

Conventional fluorophores such as fluorescein and rhodamine have found little application in immunoassays of this type since they cannot be detected sensitively enough. Conventional fluorimetric measurements are made in the presence of scattered incident light which limits the sensitivity with which the fluorescence emission can be detected. Furthermore, biological specimens contain a host of naturally fluorescent compounds which interfere with the measurement of specific fluorescence from the label. These problems have been overcome to a certain extent by the use of pulsed-light, time-resolved fluorescence measurements [15]. This requires the use of labels based on rare-earth chelates (*Figure 5.12*) such as organo-metallic coordination complexes of europium. These have two major differences compared with conventional fluorophores. Firstly, they have a very large Stokes' shift; that is, the wavelengths of the exciting and emitted radiation do not overlap hence allowing efficient optical separation. Secondly, the fluorescence lifetime is significantly greater than that of endogenous fluorophores. This latter property can be exploited to minimize interference by using a short pulse of exciting radiation, waiting for the interfering signal to decay (nanoseconds) and measuring only the remaining signal due to chelate fluorescence (microseconds). These excitation/emission cycles can be repeated many times per second in order to increase the signal. This system requires sophisticated equipment and chemistry to yield sensitive immunoassays.

5.5.3 Chemiluminescence immunoassay

Chemiluminescent end-points have been discussed in Section 5.5.1 in the context of enzyme immunoassay. However, much work has been done over the years towards exploiting the possibility of using chemiluminescent molecules themselves as the labels. Potentially, this yields simpler, more robust assays.

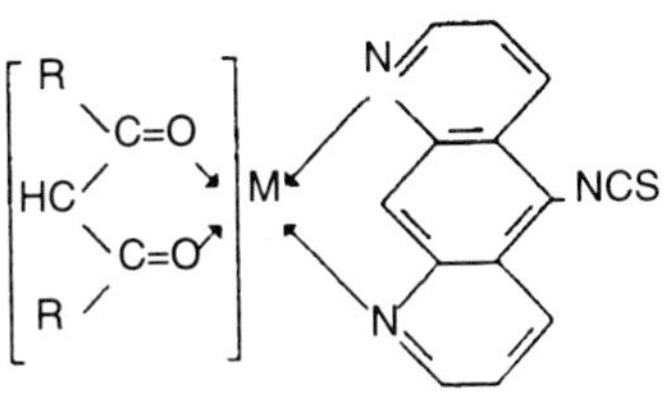

FIGURE 5.12: *Example of a typical rare-earth chelate for fluorescence labeling. M, europium: the most widely used lanthanide.*

Chemiluminescence is observed when a highly energetic chemical reaction produces product molecules in an electronically excited state. These excited molecules then relax to their ground state with the emission of photons. Relatively few classes of chemiluminescent molecules have been successfully coupled to antigens or antibodies and then used as a basis for immunoassay techniques [16]. One of the problems is the need to perform the coupling reaction without adversely affecting the quantum yield of the chemiluminescent molecule. Early work in which luminol was used to label antibodies demonstrated substantial losses of light output during reagent preparation. This was improved to some extent by the use of isoluminol derivatives which were less influenced by structural changes, but it still proved difficult to produce chemiluminescent immunoassays which could be used routinely. Later work on isoluminol derivatives and the corresponding catalysts required to bring about chemiluminescence went some way to demonstrating the routine feasibility of chemiluminescent assays.

A different approach to that described above involved the use of a class of compounds known as the acridinium salts (*Figure 5.13*). These compounds undergo chemiluminescent reactions without any need for the catalysts which are required by the luminol-type compounds. These catalysts comprise a wide variety of species from simple transition metal cations to complex molecules such as enzymes. This in turn renders them susceptible to interference and requires complex reagent systems to initiate the chemiluminescent

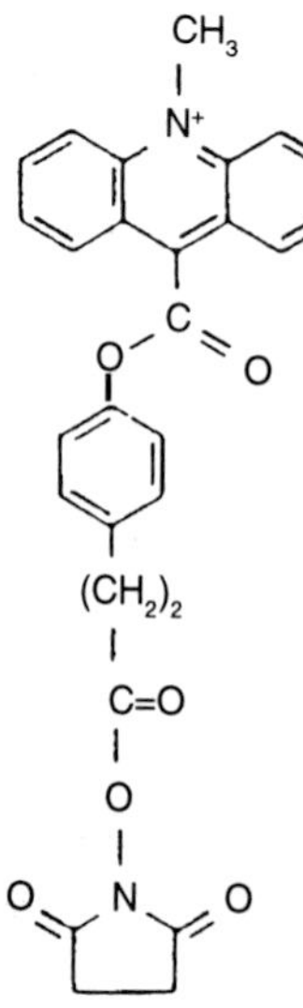

FIGURE 5.13: *A chemiluminescent acridinium ester for producing labeled immunoassay reagents.*

reaction. By contrast, the acridinium salts require addition of dilute alkaline hydrogen peroxide only to initiate chemiluminescence. Furthermore, the acridinium salts have been coupled to antigens and antibodies without loss of quantum yield [17]. Chemiluminescence instrumentation has also become more diverse in recent years and it is now possible to obtain a range of luminometers from simple portable instruments to complete random access analyzers.

5.5.4 Homogeneous immunoassays

In general, immunoassays based on radioactive labels require that the bound and free fractions are separated in order to assess the radioactivity bound in the form of immune complexes. This is necessary because the emission from the radioisotope is unaffected by its environment in this context. However, certain nonradioactive labels often have properties that are affected by their environment; in this case, whether or not they are bound in the form of immune complexes. Thus, a change in the nature or magnitude of the signal characteristic of the label is a measure of the extent of immune complex formation which is detectable or quantifiable without the prior separation of the immune complexes.

These changes are often small such that they cannot be detected with great sensitivity. This may be due to a number of factors. One particularly important consideration is that of potential interference of the label signal by the analytical matrix. In heterogeneous immunoassays, the solid-phase immune complexes are separated and often washed to maximize signal discrimination between bound and free phases. This also has the effect that potentially interfering substances are removed prior to quantitation. Thus, homogeneous immunoassays have only been of use in situations where high levels of sensitivity are not required. Such assays are, however, generally easy to use and automate. One of the most successful examples of this class of immunoassays is the enzyme-multiplied immunoassay technique (EMIT) [18] which utilizes enzyme-labeled antigen conjugates. In its free form, the enzyme conjugate will catalyze the conversion of a colorless substrate to a colored product whereas this reaction is inhibited when the conjugate is bound by antibody (*Figure 5.14*). The intensity of color produced is thus directly proportional to the amount of competing antigen in the added sample.

A further homogeneous immunoassay system to achieve popularity is the TDX system [19] which relies on the use of fluorescence polarization as an end-point. Here, a conventional fluorescent antigen conjugate is used in conjunction with a fluorimeter which utilizes

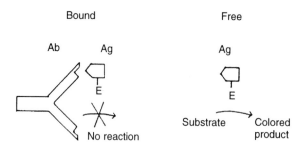

FIGURE 5.14: *Schematic representation of the EMIT used widely for homogeneous (nonseparation) immunoassays.*

plane-polarized incident light as an excitation source. The degree of depolarization of the emitted light relative to the incident light is related to the speed of rotation of the conjugate molecule. Provided that the antigen is small, the rate of rotational motion is substantially reduced when it is bound by a large antibody molecule. This loss of polarization can be quantified and related to the degree of immune complex formation with the conjugate, and hence is a measure of the amount of analyte antigen which competes for antibody binding sites in the system. Whilst many other types of homogeneous assay have been reported, none have achieved the widespread use demonstrated by this and the EMIT system.

5.6 Summary

In this chapter we have seen that antibodies can react *in vitro* with their corresponding antigens to yield immunoprecipitates. This phenomenon can be used as the basis of techniques to detect and quantify antigens or antibodies. In their simplest form these techniques are limited in their usefulness by their relatively poor sensitivity. High sensitivity is ultimately obtained by the use of immunoassays, particularly in situations where the antibody itself is labeled with a radioisotope. Increased performance can be obtained where the radioactive label is replaced by a nonradioactive alternative such as a chemiluminescent molecule.

References

1. Catty, D. and Raykundalia, C. (1988) in *Antibodies: a Practical Approach* (D. Catty, ed.). IRL Press, Oxford, p. 137.
2. Ling, N.R. and Catty, D. (1988) in *Antibodies: a Practical Approach* (D. Catty, ed.). IRL Press, Oxford, p. 169.
3. Yalow, R.S. and Berson, S.A. (1960) *J. Clin. Invest.,* **39,** 1157.
4. Hunter, W.M. and Greenwod, F.C. (1962) *Nature,* **194,** 495.
5. Corrie, J.E.T. (1983) in *Immunoassays for Clinical Chemistry* (W.M. Hunter and J.E.T. Corrie, eds). Churchill Livingstone, Edinburgh, p. 353.
6. Bolton, A.E. and Hunter, W.M. (1973) *Biochem. J.,* **133,** 529.
7. Miles, L.E.M. and Hales, C.N. (1968) *Nature,* **219,** 186.
8. Woodhead, J.S., Addison, G.M. and Hales, C.N. (1974) *Br. Med. Bull.,* **30,** 44.
9. Kohler, G. and Milstein, C. (1975) *Nature,* **256,** 495.
10. Engvall, E. and Pearlmann, P. (1971) *Immunochemistry,* **8,** 871.
11. Stanley, C.J., Johanssen, A. and Self, C.H. (1985) *J. Immunol. Methods,* **83,** 89.
12. Posch, N.A., Wells, A.F. and Tenoso, H.J. (1973) *PB-224875.* National Technical Information Service, USA.
13. Thorpe, G.H.G. and Kricka, L.J. (1986) *Methods Enzymol.,* **133,** 331.
14. Bronstein, I.,Voyta, J.C., Thorpe, G.H.G., Kricka, L.J. and Armstrong, G. (1989) *Clin. Chem.,* **35,** 1441.
15. Hemmila, I., Dakubu, S., Mukkala, V.M., Sutari, H. and Lovgren, T. (1983) *Anal. Biochem.,* **137,** 335.
16. Weeks, I. (1992) in *Comprehensive Analytical Chemistry,* Vol. 29, *Chemiluminescence Immunoassay* (G. Svehla, ed). Elsevier, Amsterdam.
17. Weeks, I., Beheshti, I., McCapra, F., Campbell, A.K. and Woodhead, J.S. (1983) *Clin. Chem.,* **29,** 1474.
18. Rubenstein, K.E., Schneider, R.S. and Ullman, E.F. (1972) *Biochem. Biophys. Res. Commun.,* **47,** 846.
19. Dandliker, W.B., Kelly, J.K. Dandliker, J. and Levin, J. (1973) *Immunochemistry,* **10,** 219.

6 Immunolocalization

6.1 Introduction

Chapter 5 describes how antibodies can be used to detect and quantify antigens in solution. Such assays, however, are rarely able to give information on the original location of the antigen in relation to its role *in vivo*. The major part of this chapter introduces the technique of immunocytochemistry in which antigens can be localized within cells and tissues by the application of antibodies specific for the antigen under investigation. The antigen–antibody complex can then be visualized by the application of various labels or markers. The different options for specimen preparation, both for light and electron microscopy, are described, together with the choices of antibody and label most appropriate for different applications, including the more complex multiple labeling strategies.

Immunoblotting, a technique for studying molecular size, is also included since the visualization steps of antibody and label application are similar. Since it is related to radioimmunotherapy, immunolocalization *in vivo* will be discussed in Chapter 7.

6.2 Immunocytochemistry

The techniques of immunocytochemistry are very simple to perform, once they have been established for a particular tissue and antigen. Establishment of these techniques for new systems, however, requires thorough optimization of each step in the process, including trials of alternative methods. The important points to consider and common choices at each step are described below, followed by descriptions of the various assay formats. Detailed technical protocols can be found in refs 1–4.

Essentially, the tissue has to be carefully prepared to preserve its morphology and is then cut into very thin sections (~ 5 µm) which are attached to microscope slides (usually pre-coated with amino-propyltriethoxysilane (APES) to aid attachment of tissue). There then follows a series of short (i.e 30-min) incubations at room temperature with solutions of primary antibody specific to the antigen under study and the various label reagents as will be described below. These incubations with active reagents are interspersed with thorough but gentle washing in a neutral buffer to wash away surplus unbound reagents. The end result is a colored insoluble product deposited in the area where the antigen is located within the tissue or cell. The result is usually only qualitative, although modern microscopy techniques are now permitting some degree of quantitation.

6.2.1 Tissue and cell preparation

The preparation of the tissue for microscopical analysis is critically important, not only to maintain the morphological structure of the sample, but to retain the antigen in its natural location within the tissue and in its natural conformation. To this end, the tissue can be fixed by immersion in a variety of different solutions which stabilize and protect the tissue from subsequent physical and chemical treatments. Fixatives act either by cross-linking between reactive groups in polypeptide chains of proteins or by disrupting hydrogen bonds. One of the most widely used groups of fixatives are formalin-based (i.e. 10% formalin in tap water, formal saline, or neutral buffered formalin) but there are also special formulations such as mercuric chloride and periodate-lysine-paraformaldehyde that have been developed for specific tissues.

Unfortunately, these preservation methods are likely to damage the antigen structure such that antibody binding is diminished or even eliminated. The solution to the problem is often a compromise between retention of morphology and maximizing antibody binding. Methods appropriate for many mammalian tissues and antigens have been established so these should form the basis of any new study, but it would be advisable initially to try several different fixation methods in order to obtain the best results. The exposure of antigens in formalin-fixed tissue to subsequently added antibodies can sometimes be improved by the application of proteolytic enzymes such as trypsin, chymotrypsin, pepsin and proteinase K. These act by breaking the protein cross-links formed by formalin, thereby exposing more antigenic sites.

After fixing, the tissue needs to be embedded in a rigid medium to facilitate cutting into thin sections. Immunocytochemistry is most often carried out in histopathology laboratories on formalin-fixed, paraffin wax-embedded tissue. This allows for long-term storage of the material with good retention of morphology. After sections are cut, the wax has to be removed with xylene, which is washed off with a series of alcohol solutions of diminishing concentration, prior to following the immunocytochemical procedures. These processes can also contribute to the destruction of antigen structure.

Epoxy and acrylic resins have also been used for embedding tissue to be used for histological staining and offer several advantages over paraffin embedding in terms of better resolution. However, such resins are not widely used for immunocytochemistry because the various technical procedures impede access to or destroy antigenic sites.

In research laboratories, immunocytochemistry is often carried out on frozen tissue that has not been fixed. Small pieces of tissue are snap frozen in isopentane immersed in liquid nitrogen, and then thin (5 μm) sections are cut on a microtome at –20°C and air dried on to slides. Morphology can be slightly poorer and samples cannot be stored as long as fixed samples, but frozen sections allow for the best retention of antigenicity.

Recently, heat treatment of paraffin sections, by microwaving or pressure cooking, has been shown to reveal previously hidden antigenic sites, surprisingly without damage to tissue morphology [5]. Several suggestions for the mechanism of action have been made, including alteration of protein tertiary structure and removal of methyl groups.

Mammalian cell suspensions do not need to be frozen or embedded in wax but can be smeared on to microscope slides as a single cell layer or centrifuged on to slides by means of a cytospin such that cells are evenly dispersed by the centifugal force.

Whole plant cells are particularly impenetrable for immuno-cytochemical staining due to the presence of rigid glycan cell walls and other intracellular structures and molecules which might interfere with subsequent procedures. Also, desiccated seeds, spores and pollens present special penetration problems. The preferred approach is to fix and embed plant material in resin before immunolabeling on the cut surface. There is no standard fixation procedure and it will vary according to the type of plant material.

The preparation of tissue and cells for electron microscopy is particularly important since the resolution achieved is so much higher than for light microscopy, so any deficiencies in preservation of tissue morphology will be more evident. The problem remains, however, that the better the preservation of tissue morphology, the more likely it is that antigen structure is impaired and penetration of the tissue by antibodies or antibody–enzyme conjugates is reduced. For these reasons, pre-embedding staining procedures have been developed in which permeability of tissue is maximized, the antibody is bound to antigen and the subsequent labeling is carried out before the tissue is fully fixed, embedded in resin and sectioned. Alternatively, antigen detection is carried out on ultrathin sections (< 0.1 μm) after embedding and sectioning. The usual fixatives are glutaraldehyde or formaldehyde based, but detailed procedures should be followed from specialist texts [1,2,6].

6.2.2 Choice of antibody

Obviously, the choice of antibody will be influenced by the availability of antibody specific for the antigen under study. Many of these, especially in the area of human diagnostics, will be available commercially. Within that category, however, one may have many choices (e.g. polyclonal, monoclonal, serum, ascites, culture supernatant, IgG, IgM, Fab fragments, labeled or unlabeled) and the cost can vary considerably.

Polyclonal antisera will be the cheapest option (and easiest to make oneself) but present the greatest risk of cross-reaction with nontarget antigens, not only with respect to the binding specificities of the constituent antibodies but also the 'unknown' constituents of the serum, which may bind to the tissue and subsequently to the labeling reagents to give false positives. However, if the antiserum has been affinity purified and carefully characterized, its binding to many epitopes on a specific antigen will be an advantage in immuno-cytochemistry and will help to counteract the destructive effects the fixative may have on individual epitopes, i.e. there is less likelihood of all epitopes being destroyed than just one or two.

Monoclonal antibodies, of course, can offer unique specificity but if that epitope is liable to damage during the fixative procedures, the monoclonal antibody will not bind at all. Also, if the epitope is not abundant, using one monoclonal antibody may not be sufficiently sensitive to reveal the antigen. However, using a cocktail of monoclonal antibodies of different specificities is a possible option.

Given the choice, a pure immunoglobulin fraction is preferable to serum, especially if the immunoglobulin is affinity purified. Ascites, however, can be used without purification since it can usually be used at such a high dilution that other components in the fluid become insignificant. IgG is the preferred fraction since it usually has a higher affinity for antigen, and is easier to purify (and label directly) than IgM. However, if an IgM of reasonable affinity is available, the pentameric structure can provide significant signal enhancement when used in conjunction with an anti-IgM label (see Section 6.2.4, p. 94), although the larger size of the complex may cause steric hindrance and reduced penetration of tissue. Fab fragments are smaller than whole antibodies, which may be an important consideration where good penetration of tissue is required, but it may also be necessary to use Fabs if the specimen contains Fc receptors which will bind the antibody 'nonspecifically'.

6.2.3 Choice of antibody label

Antibody labels for immunocytochemistry either fluoresce when exposed to ultraviolet (UV) light or deposit an insoluble product at the site of the antigen–antibody complex. The latter are usually enzymes conjugated to the antibody which, when exposed to the enzyme substrate, produce a colored product. Alternatively, electron dense gold particles can be attached directly to the antibody.

Fluorescent labels. The earliest antibody label to be used in immunocytochemistry was the fluorescent marker fluorescein isothiocyanate (FITC), and it is still a popular choice today. When the tissue is exposed to UV light of a specific wavelength (excitation), the FITC will emit a fluorescent apple-green light through barrier filters of the appropriate wavelength (see *Table 6.1*). Many antibodies can be purchased already labeled with FITC but the procedure of labeling one's own antibody is very simple and cheap, requiring little more than a few hours incubation of the purified antibody and fluorochrome followed by gel filtration to separate the conjugate from unreacted substrates. A fluorescent microscope with the correct excitation and barrier filters and a camera attachment is required. The image is prone to rapid fading (or quenching) under exposure to UV light, although there are special slide mountants available that can help to counteract this effect such as diazabicyclo[2.2.2]octane (DABCO). Nevertheless, fluorescent signals can be very sensitive and the pattern of fluorescence (i.e. fibrillar, particulate or diffuse) can provide characteristic diagnostic clues. However, the tissue background detail is poor compared with that obtainable with enzyme labels, although it is possible to counterstain to some extent. The DNA stain 4′, 6-diamidino-2-phenylindole (DAPI), which is excited by UV light has a blue emission wavelength and so is a suitable background for red or

TABLE 6.1: Common fluorochromes used in immunofluorescence

Fluorochrome	Acronym	Excitation wavelength (nm)	Emission wavelength	Color
Fluorescein isothiocyanate	FITC	495	515	Apple-green
Tetramethylrhodamine isothiocyanate	TRITC	575	600	Red
Texas red		595	615	Red
7-amino-4-methyl-coumarin-3-acetic acid	AMCA	350	450	Blue
Phycoerythrin	PE	450–470	575	Orange
Cyanins	Cy5 Cy3	648 554	665 568–574	Red Orange

green specific labels. Propidium iodide (PI) gives a red background under the same excitation as is used for FITC specific staining.

Other fluorochromes suitable for antibody labeling which can emit different colored signals, together with their specific excitation and emission wavelengths, are listed in *Table 6.1*. The available filter combinations have improved significantly over the last decade and advice as to which are the most suitable for a particular application should be sought from the microscope manufacturers, especially with regard to visualizing different colors separately and simultaneously.

Several new fluorochromes have been introduced which facilitate visualization of several antigens (and therefore colors) simultaneously. Phycoerythrin, for example, has a broad excitation range which enables it to be excited at the same wavelength as FITC thus enabling orange and green to be seen simultaneously. The cyanins have also improved visualization since they tend to be more hydrophilic than other fluorochromes so there is less of a tendency to aggregate and cause nonspecific binding. They are also more photostable so the quenching effect seen with FITC is minimal. Cy3 can be excited by either blue or green light to give orange or red signals as required.

Visualizing whole cells or layers with a conventional fluorescence microscope can result in some distortion of the image as a result of light interference from the out-of-focus layers. The confocal scanning optical microscope overcomes this problem since it only receives light from the in-focus image. By taking a series of optical sections through the material, a picture of the three-dimensional distribution of light can be built up through computer imagery.

Enzyme labels. There are four enzymes in common use for immuno-cytochemistry, peroxidase, alkaline phosphatase, glucose oxidase and microperoxidase, and many antibodies can be obtained commercially already conjugated to at least the first two. Of these, peroxidase (hydrogen-peroxide oxidoreductase) is the most popular for use on mammalian tissue. It is widely distributed in plant tissues, but less so in animal tissues, as different isozymes. The commonest form for this application is the 'C' isozyme of horseradish. Its function is to transfer hydrogen ions from hydrogen donors to H_2O_2 to make water. There are numerous hydrogen donors that are suitable substrates for immunocytochemistry, i.e. those that produce a colored insoluble product (see *Table 6.2*). The choice of substrate is mainly dependent on the desired color and the likelihood of the substrate being converted by endogenous enzymes in the specimen (which is why alternative enzymes are preferred for plant tissue). Some enzyme substrates such as 3,3'-diaminobenzidine (DAB) are potentially carcinogenic, so should be handled with care and preferably bought in single-use tablet form rather than as loose powder. DAB end-products can be amplified by as much as 100-fold if required by application of the silver or nickel enhancement techniques (see p. 93).

TABLE 6.2:   *Common enzymes used in immunocytochemistry and their substrates*

Enzyme	Substrate	Color of product
Horseradish peroxidase (HRP)	DAB (3,3'-diaminobenzidine)	Brown
	4-CN (4-chloro-1-naphthol)	Blue-black
	AEC (3-amino-9-ethylcarbazole)	Reddish
	Hanker–Yates reagent (*p*-phenylenediamine-HCl and pyrocatechol)	Purple-brown
	TMB (5,5'- tetramethylbenzidine)	Dark-blue
	α-naphthol/pyronin	Red-purple
Alkaline phosphatase (AP)	Naphthol AS phosphate +	
	Fast red TR	Red
	Fast blue BBN	Blue
	Fast red violet LB	Violet
	Bromochloroindolyl phosphate (BCIP) and nitroblue tetrazolium (NBT)	Blue
Glucose oxidase (GO)	*t*-NBT and *m*-PMS (nitroblue tetrazolium chloride and phenazine methosulfate)	Blue-purple

A popular counterstain for use with DAB is hemotoxylin which stains nuclei and basophilic cytoplasms blue contrasting well with the brown of DAB. However, there are many different colored stains which react with specific cellular molecules to aid morphological identification [3].

Alkaline phosphatase (AP) obtained from intestinal tissue or bacteria is often preferred for staining plant tissue since it is absent from higher plants but is unsuitable for staining intestinal and some other mammalian tissues because of endogenous enzyme. The properties of AP from intestinal or bacterial sources are very different. Bacterial AP has a lower activity but has a lower pH optimum (approximately pH 8) compared to that of the intestinal enzyme (approximately pH 10). The alkaline buffers used for intestinal AP staining often cause dissociation of antigen and antibody resulting in diffuse staining. Thus, bacterial AP often gives better results at the near-neutral pH values required to maintain antigen–antibody binding. AP hydrolyzes phosphate esters, such as alcohols, phenols and amines, and so is very susceptible to the buffers used. For example, a common wash buffer used in immunocytochemistry is phosphate-buffered saline, but it contains sufficient inorganic phosphate to inhibit AP competitively. Tris buffers should be used in preference. The usual substrate for AP in immunocytochemistry is naphthol AS phosphate; liberated naphthol can then be combined with fast red, blue or violet according to which color is preferred (see *Table 6.2*). Fast red contrasts well with hematoxylin counter stain.

Glucose oxidase (β-D-glucose:oxygen-1-oxidoreductase) from fungal sources is often favored for its very low background staining because endogenous activity is absent from mammalian tissue. With its substrate, nitroblue tetrazolium and phenazine methosulfate, a blue-purple insoluble product is formed.

Microperoxidase (MPO), a proteolytic fragment of cytochrome c from horse heart, has a molecular weight of only 1500–2000. It is used only when penetration of bulky antibody–enzyme conjugates into tissues presents a problem, and almost exclusively in electron microscopy. Approximately 20–50 MPO fragments can be attached to an IgG or Fab molecule before diminishing antibody binding by steric hindrance. Thus, although the activity of MPO is poorer than horseradish peroxidase (HRP), because of this high conjugation ratio, antibody–MPO conjugates are only slightly less active than antibody–HRP conjugates and are only half the diameter.

Metal labels. Colloidal gold is a useful antibody label at both the light microscope level and, for finer resolution, in electron microscopy. Antibodies are labeled with gold particles by noncovalent electrostatic adsorption in a relatively simple process once all variable factors, such as concentrations and pH, are optimized. The label can be visualized as a red coloration but the signal can be amplified with the silver enhancement technique to produce a black electron-dense reaction product. Metallic gold catalyzes the reduction of silver ions to metallic silver in the presence of a reducing agent; thus, the gold particles on the labeling reagent become coated with visible metallic silver. In a similar way, silver can be used to enhance the reaction between HRP and its substrate DAB. DAB staining can also be enhanced by nickel. Colloidal silver has been used alone and gives a yellow signal.

Colloidal gold is the probe of choice for electron microscopy. Its advantages are that it can be seen as a discrete spherical particle that can easily be distinguished from cellular morphology, gives the highest resolution of any probe, is independent of any endogenous enzyme reactions and can even be counted (see *Figure 6.1*). The

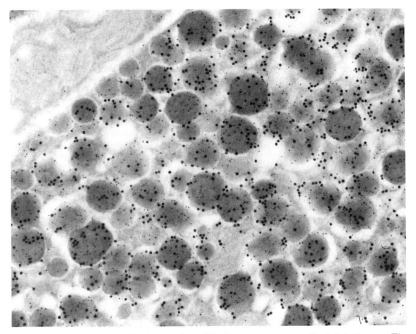

FIGURE 6.1: α-granules in an A cell in human pancreatic islet tissue. The tissue was fixed in neutral buffered 1% glutaraldehyde and embedded in the acrylic resin LR White. Thin sections were immunolabeled first with rabbit anti-glicentin and goat anti-rabbit Ig linked to 20 nm colloidal gold, then with hapten-labeled rabbit anti-glucagon followed by goat anti-hapten Ig linked to 10 nm colloidal gold. X 42 000. Photograph courtesy of Dr G.R. Newman, EM Unit, University of Wales College of Medicine, Cardiff, UK.

particles are available in a range of sizes (1–40 nm) which, when used to label different probes, can distinguish between different epitope locations very clearly without problems of penetration or steric hindrance. These advantages of gold probes have resulted in the replacement of ferritin and hemocyanin that were once used.

6.2.4 Labeling strategies

Direct labeling methods. The steps described in the previous sections imply a simple staining protocol in which the tissue is prepared and the specific antibody is pre-labeled with enzyme or fluorochrome, and then incubated with the tissue antigen and visualized under UV light or by application of the enzyme substrate (*Figure 6.2a,b*). Whereas such a scheme is possible, it is rarely used in practice. Unless a particular antibody is to be used frequently, it is not cost effective to label each specific antibody preparation, especially since purification of the immunoglobulin is required. It is also unlikely to be as sensitive as the alternative indirect labeling schemes which provide better opportunities for improving sensitivity by increasing signals and reducing background staining.

Indirect labeling methods. The simplest indirect labeling method allows the use of unlabeled primary specific antibody at the first step, and the signal is provided by binding of an anti-species immuno-globulin conjugated with enzyme, fluorochrome or metal (see *Figure 6.2c*). For example, if the primary antibody is derived from a rabbit, the secondary labeled antibody will be an anti-rabbit IgG (or IgM) raised in another species such as goat or sheep; if the primary antibody is a mouse monoclonal Fab fragment, the secondary antibody will be an anti-mouse Fab conjugate. This system allows great flexibility since any primary antibody of the same species can be used with one anti-species antibody label conjugate, which can reduce costs considerably. In addition, several molecules of the secondary antibody and, therefore, label can bind to the primary antibody as long as steric hindrance does not interfere, thus enhancing the signal-to-antigen ratio. Care should be exercised, however, when using multiple species (including the species of the tissue in question) since nonspecific binding can occur.

An alternative nonantibody indirect label takes advantage of the property of Protein A to bind to the Fc part of an antibody. Protein A is conjugated to the label and is applied in the same way as the secondary antibody label described in the previous paragraph (see *Figure 6.2d*). One major problem with this technique is that it does not work with some antibodies, since Protein A does not bind to all

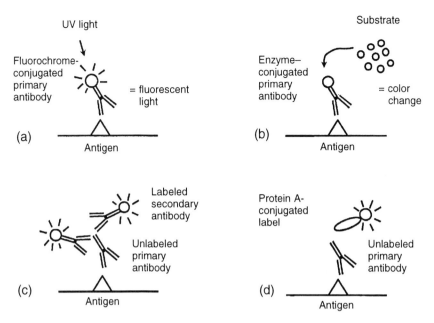

FIGURE 6.2: *Simple immunocytochemical staining procedures. (**a**) Direct labeling: primary antibody pre-labeled with FITC. A fluorescent green signal is emitted on exposure to UV light. (**b**) Direct labeling: primary antibody pre-labeled with enzyme. A colored insoluble product is deposited at the antigen site on exposure to enzyme substrate. (**c**) Indirect labeling: primary unlabeled rabbit antibody bound to antigen, followed by anti-rabbit IgG antibody conjugated to label (or similarly for other species). (**d**) Indirect labeling: primary unlabeled rabbit antibody bound to antigen, followed by Protein A conjugated to label.*

immunoglobulin isotypes or species, nor with equivalent avidity. Protein A can also bind to certain Fc receptors and cause 'nonspecific binding'.

Another nonantibody indirect labeling method is the avidin–biotin system. Avidin is a basic glycoprotein of 68 kDa which has an exceptionally high affinity for the vitamin biotin (244 Da). The affinity constant (K) is of the order of 10^{15} M^{-1}, which represents much stronger binding than the best binding affinities of antibodies with their antigens. This property can be exploited for labeling proteins, especially since biotin is very easily conjugated to primary amino groups on most proteins, including antibodies and enzymes. Biotin is obtainable in the form of biotinyl-N-hydroxysuccinimide ester, the activated ester group of which binds to the nucleophilic unprotonated ε-amino groups of lysine on a protein (i.e. antibody) to form an amide bond. Although there are four binding sites on avidin for biotin, in practice these are paired, so avidin is best used as a bridging molecule between a biotinylated primary antibody and a biotinylated enzyme. However, some enzymes, such as AP, are inactivated by biotin, so avidin should be bound directly to the enzyme under these circumstances (see *Figure 6.3a*).

There are some problems with the use of avidin due to its high isoelectric point (approximately 10) with the result that it is positively charged at neutral pH and, therefore, liable to bind to negatively charged structures such as nuclei giving rise to high nonspecific binding. In addition, it is a glycoprotein and can bind to lectins via the carbohydrate moiety. These problems can easily be overcome, however, by using streptavidin (sometimes known as extravidin), a 60-kDa protein from *Streptomyces avidinii* with similar high affinity for biotin, which is not strongly charged at neutral pH and is not a glycoprotein. Streptavidin is available commercially, bound to a number of enzymes and fluorochromes, and also magnetic beads for cell separation and protein purification requirements. Using the biotin–streptavidin system, rather than secondary antibody labels, avoids problems that might arise over nonspecific binding of the secondary antibody to the tissue. Additionally, if several molecules of biotin are bound to the antibody (usually with spacer arms to avoid steric hindrance), the binding of streptavidin either directly labeled or with biotinylated enzymes will have a signal amplification effect.

A popular way of amplifying the amount of label at the antigen site is to apply pre-formed complexes of three molecules of enzyme bound to two molecules of specific anti-enzyme antibody. Examples are peroxidase–antiperoxidase (PAP) and AP–anti-AP (APAAP), and are illustrated in *Figure 6.3 (b,c)*. The primary unlabeled antibody is first allowed to bind to antigen, followed by an anti-rabbit or -mouse IgG according to the species of the first antibody. This leaves one remaining arm of the anti-IgG to bind to an IgG of the enzyme complex; thus, a bridge is formed with three molecules of enzyme attached, to react with the substrate. This has been shown to be a very sensitive technique.

6.2.5 Double and triple labeling

All the different labeling methods described in the previous section are very useful when more than one antigen needs to be visualized in a tissue sample at the same time. It would not be possible, for example, to use two mouse monoclonal antibodies as primary antibodies, followed by the same anti-mouse IgG label, since the label would bind to both primary antibodies and one would be unable to

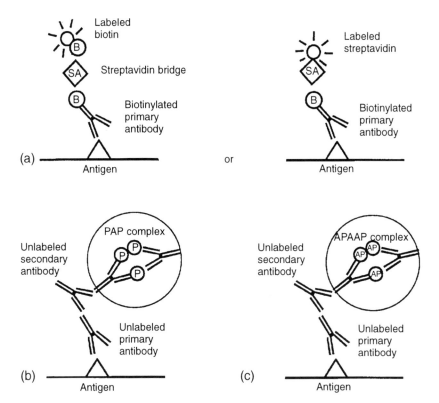

FIGURE 6.3: *Label complexes. (a) Biotin (B)–streptavidin (SA) system. (b) Peroxidase (P)–anti-peroxidase (PAP) system. If the primary antibody is of mouse origin, the secondary antibody is an anti-mouse IgG which will also bind to the pre-formed mouse PAP complex. Alternatively, rabbit primary antibody/anti-rabbit IgG/rabbit PAP can be used. (c) AP–anti-AP (APAAP) system. As (b) but the enzyme is AP.*

distinguish between the two (but see *Figure 6.4c*). Similarly, if one primary antibody was of mouse origin and the other rabbit, one could use an anti-mouse IgG label and an anti-rabbit IgG label but, if both were conjugated with HRP, again the two antigens could not be distinguished. Compatible secondary antibody–enzyme conjugates and substrates would be peroxidase with DAB substrate (red) and AP with fast blue for contrast. Alternatively, one of the primary antibodies could be directly labeled and the other could be biotinylated with an alternative enzyme conjugated to streptavidin. Some examples of the possible labeling sequences are given in *Figure 6.4* but careful attention to possible nonspecific interactions between reagents with adequate blocking of nonspecific sites, negative controls and thorough washing between steps is even more important with these complex schemes (see Section 6.2.6).

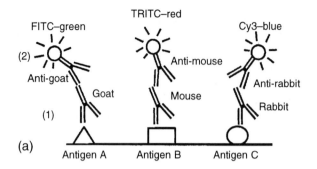

(a)

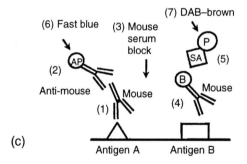

(b)

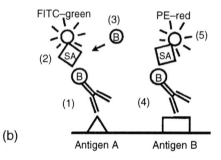

(c)

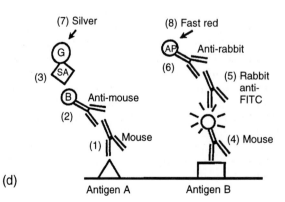

(d)

Multiple labeling with enzymes can still create problems of color mixing, especially if the antigenic sites are close together and at very different concentrations. Fluorescent labels, however, do not have this drawback, since different antigens can be visualized either separately or simultaneously. Red, green and blue signals can easily be detected simultaneously if the appropriate combination of UV filters is used. FITC and the relatively new fluorochrome phycoerythrin can both be seen under the same excitation wavelength and emit green and orange, respectively.

In electron microscopy, multiple antigen labeling is easily achieved by conjugating each specific antibody preparation with gold particles of different sizes (see *Figure 6.1*).

6.2.6 Nonspecific staining

Nonspecific staining can be a problem in immunocytochemistry if sufficient care is not taken, both in the choice of staining method and its implementation. Most of these problems can be overcome by systematically optimizing each step, taking certain precautions or adopting one of the alternative labeling schemes outlined above.

The problems can be due to specific cross-reaction of the primary or secondary antibodies with common epitopes on other antigens. These· problems can be overcome by further purification of the antibodies or by using alternative antigen-specific antibodies. Some so-called species-specific antibodies may cross-react with other species and should be checked before use. The Fc part of antibodies may also

---

FIGURE 6.4:   *Multiple labeling strategies. (**a**) (1) A mixture of primary antibodies specific for three different antigens is added, all raised in different species. (2) A mixture of the three corresponding anti-species IgG antibodies is then added, each conjugated to a different colored fluorochrome. (**b**) (1) Biotinylated mouse primary antibody to antigen A. (2) Streptavidin (SA)–FITC label. (3) Remove excess streptavidin label by blocking with free biotin (B). (4) Second primary biotinylated antibody to antigen B. (5) Streptavidin–phycoerythrin label. (**c**) (1) Mouse antibody to antigen A. (2) AP-labeled anti-mouse IgG. (3) Block with mouse serum. (4) Biotinylated mouse antibody to antigen B. (5) Streptavidin–peroxidase (P) label. (6) Fast blue substrate for AP. (7) DAB substrate for peroxidase. (**d**) (1) Mouse antibody to antigen A. (2) Biotinylated anti-mouse IgG. (3) Streptavidin–gold (G). (4) FITC-labeled mouse antibody to antigen B. (5) Rabbit anti-FITC. (6) AP-labeled anti-rabbit IgG. (7) Silver enhancement of gold. (8) Fast red substrate for AP. See* Table 6.1 *for fluorochrome acronyms in full.*

cause problems by binding to Fc receptors which are found on the surface of a variety of cell types. Similarly, Protein A may bind to certain Fc receptors. This can be overcome by blocking the receptors with irrelevant immunoglobulin or by applying antibody (Fab) from which the Fc portion has been removed. Alternatively, further dilution of the antibody or reducing incubation times may reduce the nonspecific binding to insignificant levels. In unfixed whole cells, immune complexes of free antigen and antibody, or primary and secondary antibodies, can be formed through cross-linkage, especially between bi- or multivalent antibodies. These form concentrated patches which accumulate at one pole of the cell, hence the term capping. The phenomenon can be minimized by using monoclonal antibodies or Fab fragments.

Some tissues are likely to contain sufficient endogenous enzyme to react significantly with added substrate. Endogenous peroxidase is common in erythrocytes, neutrophils and macrophages, and can easily be blocked in most situations by application of periodate, borohydride or sodium nitroferricyanide phenyl hydrazine. Endogenous AP in placenta can be blocked with levamisole or in intestine with 20% acetic acid. Alternatively, a different labeling system could be used. Some tissue structures may autofluoresce, giving a diffuse yellow-green coloration with filters used for FITC staining. This too can be overcome by pre-staining with a fluorescent background dye such as pontamine sky blue or by changing the fluorescent label.

Adverse reactions might occur between the fixative or embedding compound and the substrate solutions but there are invariably alternative reagents available. Nonspecific binding can also be caused by inadequate washing of the specimen or by allowing it to dry in between incubation steps. Precise protocols, including ways to identify and avoid nonspecific reactions, can be found in refs 1 and 2.

6.3 Immunoblotting

Gel electrophoresis is a powerful technique for the separation and analysis of mixtures of proteins. Probably the commonest form of protein electrophoresis is polyacrylamide gel electrophoresis (PAGE) in the presence of sodium dodecyl sulfate (SDS). The SDS confers a uniform charge on the protein components so that, when the proteins in the gel are separated by application of an electric current, the proteins will be separated on the basis of molecular weight. These bands can then be revealed by staining with a general protein stain such as Coomassie blue or, for more precise identification, they can be immunostained. Immunostaining is best carried out in the absence of SDS, which can interfere with enzyme reactions, so the technique of immunoblotting (sometimes known as Western blotting) was devised whereby the separated proteins on the gel are electrically transferred on to nitrocellulose paper and retain their original spatial separation [7].

Many of the labels and labeling procedures already described for immunocytochemistry also apply to the technique of immunoblotting. In addition, antibodies labeled with radioisotopes (i.e. ^{125}I) are sometimes used, especially if extra sensitivity is required. After blotting, the nitrocellulose paper simply needs to be incubated sequentially with the active reagents and washed thoroughly at each step to remove unbound material. HRP is probably the commonest enzyme used for this technique, with 4-CN as the most suitable substrate (see *Table 6.2*). Paper with radioactive labels needs to be exposed to X-ray film to reveal labeled protein bands (see *Figure 6.5*). Revelation of a specific band can be correlated with molecular weight

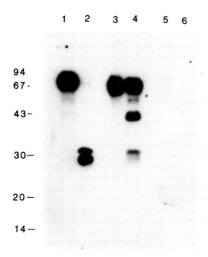

FIGURE 6.5: *An immunoblot (Western blot) following SDS–PAGE of purified clotting factor IX from different patients, visualized using a radiolabeled (^{125}I) polyclonal antiserum to factor IX and exposed to X-ray film. Odd numbers represent native factor IX protein, even numbers represent activated factor IX. Lanes 1 and 2 are protein from a normal patient, 2 and 3 protein from a patient with an abnormal factor IX gene, and 5 and 6 from a patient with a complete factor IX gene deletion. Numbers to the left represent standard molecular weight marker positions. Photograph courtesy of Dr M.B. Liddell, formerly of Dept of Haematology, University of Wales College of Medicine, Cardiff, UK.*

in comparison with standard molecular weight markers run in parallel. Multiple bands may indicate different proteins but also denatured sections of a multimeric protein. There can often be problems with monoclonal antibodies in this application since the original epitope that the MAb recognized can easily be destroyed by the denaturing effect of the SDS.

References

1. Beesley, J.E. (ed.) (1993) *Immunocytochemistry: a Practical Approach*. IRL Press, Oxford.
2. Tijssen, P. (1985) *Practice and Theory of Enzyme Immunoassays*. Elsevier, Amsterdam.
3. Horobin, R.W. (1988) *Understanding Histochemistry: Selection, Evaluation and Design of Biological Stains*. Ellis Horwood, Chichester.
4. Leitch, A.R., Schwarzacher, T., Jackson, D. and Leitch, I.J. (1994) *In Situ Hybridization*. BIOS Scientific Publishers, Oxford.
5. Cattoretti, G., Pileri, S., Parravicini, C., *et al*. (1993) *J. Pathol.* **171,** 83.
6. Polak, J.M. and Priestley, J.V. (eds) (1992) *Electron Microscopic Immunocytochemistry*. Oxford University Press, Oxford.
7. Hames, B.D. and Rickwood, D. (eds) (1990) *Gel Electrophoresis of Proteins: a Practical Approach* (2nd edn). IRL Press, Oxford.

7 Therapeutic Applications of Antibodies

7.1 Introduction

The concept of using antibodies as 'magic bullets' in the treatment of disease was first proposed by Ehrlich in 1901 (see ref. 1). This ideal has still not been fully realized in clinical practice despite the excitement aroused by the introduction of monoclonal antibodies in 1975. Polyclonal antibodies have been used for therapy for many years and, despite their undesirable side effects, they still have an important role in certain applications perhaps because of, rather than in spite of, their polyspecificity. Monoclonal antibodies can offer unique specificity for their target and therefore potentially fewer adverse reactions due to nontarget binding. However, although there is now a vast range of different specificities of rodent monoclonal antibodies, their use in human therapy has been limited mainly by the immune response of the host to the foreign protein. This response can be reduced by humanizing the original rodent antibody. With the recent developments in human recombinant antibody construction a whole new range of antibody specificities, particularly those directed at self-antigens, is now possible.

The specificity of an antibody for its target is only one of the factors to be considered before an antibody can be approved for human therapy. The effector function of the antibody is important for its ability to harness the patient's own immune response, but an antibody can also be used as a specific means of delivering a toxin, radiolabel or enzyme prodrug to its target, and also as a block of receptors or adhesion molecules. In addition, all therapeutic agents have to be thoroughly tested for their toxicity *in vitro* and *in vivo* with respect to target and nontarget cells and tissues, for viral and bacterial contamination, dose concentration and dosage regime, distribution in the body and clearance from it of both active agent and metabolites. All these considerations will be discussed in this chapter together with examples of how antibodies are currently being used therapeutically and their future potential.

7.2 Antibody type and specificity

The production and construction of the various antibody types have been described in previous chapters. Examples of how these different preparations are already being used therapeutically are given in the following sections. Technological developments are currently overtaking the relatively slow process of pre-clinical and clinical trials, so many of the newer recombinant antibodies are still being tested at the level of animal models.

7.2.1 Polyclonal antisera

Polyclonal antisera to specific targets are still useful for the treatment of acute infection such as tetanus, botulism, hepatitis and snake venom poisoning, when prompt administration of sufficient quantities of inactivating serum can save lives. In these situations, development of anti-globulin responses (see Section 7.2.2) which are seen after repeated treatments are not a major concern in comparison with the urgency of treatment. In organ transplantation, polyclonal anti-lymphocyte globulin (ALG) and anti-thymocyte globulin (ATG) are used as immunosuppressive agents, although more specific monoclonal reagents are now replacing these in many countries.

Polyclonal sera are also still in use for the prevention of newborn hemolytic disease induced by pregnancy in Rhesus-negative mothers. Such mothers are treated with sera from Rhesus-negative male volunteers who have been hyperimmunized with Rhesus D antigen. Rhesus-positive babies from subsequent pregnancies are then at less risk of developing hemolytic anemia due to the mother's immune response. Although this is an effective treatment there are problems with immunizing volunteers and clearly the human sera must be screened for human immunodeficiency virus (HIV) and other viral pathogens.

Multispecific polyclonal antisera also have a useful role therapeutically. Intravenous immunoglobulin (IVIG) is a concentrated form of IgG, also known as gamma globulin, that is derived from the pooled serum of a large number of donors. IVIG contains many types of antiviral and antibacterial antibodies. Since its introduction in 1981, IVIG therapy has become an important form of treatment in autoimmune hematologic disorders [2] and recurrent bacterial infections, for example due to B-cell chronic lymphocytic leukemia [3].

The mechanisms of action of IVIG are not fully understood in all applications, although passive replacement of antibodies to compensate for deficient circulating antibody content is a likely explanation in immunodeficiency diseases and acute infections. The immediate response to therapy of autoimmune cytopenias (cell destruction) is thought to be related to nonspecific Fc-receptor blockade of mononuclear phagocytes in the reticuloendothelial system, or in acquired coagulation disorders to idiotypic antibody interaction with pathologic autoantibodies. Although less frequent, long-term responses to IVIG therapy have been reported. Such responses must involve an immunomodulating effect of IgG that influences T- and B-cell function, with inhibition of pathological autoantibody formation. It is possible that idiotypic antibody interactions play a part in long-term responses (see Section 7.2.2, p. 108).

For general widespread use, involving more specific targeting and repeated treatments, the problems of administering polyclonal antisera lie mainly with their relative nonspecificity, the small size and variability of batches and, in the case of heterologous sera, the development of anti-antibody responses in the recipient (see Section 7.2.2). Even human sera need particularly rigorous screening to avoid transmission of hepatitis and HIV infections.

7.2.2 Rodent monoclonal antibodies

It is a common misconception that monoclonality is synonymous with monospecificity. A particular monoclonal antibody may be uniquely specific for its epitope, but that epitope may also be present in different antigens or cell types. Unexpected cross-reactions arise fairly frequently and have the potential to reduce clinical efficacy or cause pathological problems [4, 5]. Before an antibody preparation can be approved for therapy, it is a requirement that it should be thoroughly tested *in vitro* for cross-reactions with a wide range of human and animal tissue, including blood cells and cell lines, in different assay formats. However, cross-reactivity need not necessarily be a barrier to therapeutic application. Low density of epitope, anatomical barriers or poor endocytosis could prevent a nontarget cell from being killed.

Most patients will develop anti-antibodies following administration of rodent monoclonal antibodies, since they are only about 60–70% homologous with their human counterparts. The response is known as the human anti-mouse antibody (HAMA) response [6]. Some patients develop the response after a single dose and others after several doses but it usually occurs within 10-30 days of administration. Further administration of the antibody leads to the formation of immune complexes and rapid clearance to the liver and spleen, thus preventing the therapeutic antibody from reaching its target, accompanied by allergic reactions in the patient. The response can be delayed, allowing for a few more doses to be administered, by immunosuppressing the patient with drugs such as cyclosporin A which suppresses cell-mediated immunity. Using Fab fragments or humanized, CDR-grafted antibodies and new human recombinant antibodies helps to reduce the HAMA response, although any antibody conjugated to toxins or enzymes will remain immunogenic. Other strategies of avoiding or delaying the HAMA response include masking of immunogenic sites with polyethylene glycol or conjugation to daunomycin leading to cytotoxic clonal elimination of reactive B cells, or co-administration of cyclosporin A which is not without its own side effects. However, these possible adverse effects of the anti-globulin response must be kept in perspective and weighed against the benefits of treatment with the antibody.

Intact antibodies will have a longer half-life in the body than will fragments or recombinant scFvs (days compared to hours). Intact antibodies are important for prophylactic or passive immunization applications since they need to circulate for as long as possible. Small fragments (< 30 kDa) are better at penetrating solid tumors but also have faster clearance rates. This can influence the development of HAMA and the effectiveness of the antibody in reaching its target and exerting its effects. Antigen-binding fragments rather than whole antibodies may be necessary to avoid unintentional binding to Fc receptors, complement and other cells of the immune system. The Fc portion is also unnecessary when simple blocking of receptors is required. The most useful antibody specificities for therapy are those that are specific for tumor or malignant cell surface markers, those that can inactivate infectious agents and those that can manipulate the body's own immune system.

Anti-cancer antibodies. Since the development of monoclonal antibodies, a great deal of knowledge has been gained on cell-specific differentiation antigens. For example, the antigens of the major histocompatibility complex (MHC) have been classified. Some of these are useful targets for immunotherapy, either to harness the patient's own immune response more effectively or to deliver toxic agents to the target cell. CD20, for example, is an excellent antigenic target for radiotherapy as it is neither internalized nor shed after antibody binding. Antibodies used to deliver toxins to target cells are most effective if they bind to a cell surface marker that is internalized and hence transfers the toxin into the cytosol. An ideal target for this purpose is CD22.

The treatment of lymphoid malignancies has benefited particularly from immunotherapy. For example, there is a humanized mouse monoclonal antibody directed against a form of the interleukin 2 receptor (IL-2Rα) that is not present on resting T cells but is present on abnormal T cells in some leukemias and lymphomas and certain autoimmune diseases, and on effector T cells involved in graft-versus-host disease (GVHD) and allograft rejection. This antibody has been tested in Phase I clinical trials on patients with GVHD and various IL-2Rα-expressing leukemias and lymphomas. No patient developed an anti-globulin response compared with the original mouse antibody, and a significant proportion of the patients experienced partial or complete remissions [7].

The treatment of solid tumors has been less successful, due largely to the relatively poor accessibility of such tumors. Tumor size makes an important contribution to the effectiveness of immunotherapy, there being an approximate 10-fold increase in localization in tumors of less than 100 mg in comparison with larger tumors. Therefore, treatment of early disease or small metastases is likely to be more effective. Another strategy might be to use combinations of antibodies to different epitopes, such as anti-carcinoembryonic antigen (CEA) and anti-17-1A antibodies, on colorectal cancer cells which have been shown to localize more tumor cells than either alone. Other factors relating to the use of radiolabels and toxins are discussed below (pp. 112–115).

Anti-immune system antibodies. The first monoclonal antibody to be approved for therapy was OKT3 (Orthoclone) in 1986 [8]. It is a mouse monoclonal antibody (IgG2a) directed against one of the chains of the human T-lymphocyte CD3 complex of the T-cell receptor. Thus, binding to this site blocks T-cell activation and *in vivo* rapidly clears CD3-positive T cells from the peripheral circulation. Most of these T cells are not lysed but shed all their surface T-cell receptor complexes on contact with OKT3. These are replenished when OKT3 treatment is discontinued. OKT3 is an extremely useful immunosuppressive agent against T-cell mediated rejection of renal allografts, in particular, but is also being used to prevent cardiac and hepatic transplant rejection. The side effects of treatment are related to cytokine release from lymphocytes and monocytes due to T-cell activation, increased risk of viral infection due to the immunosuppression and the HAMA response. The CAMPATH series of antibodies directed against CDw52 described in Section 7.2.3 is also proving useful for immunomanipulation.

Anti-viral antibodies. Polyclonal viral neutralizing antibodies have been used for many years to reduce acute infections. The mechanism of neutralization can involve killing the virus or blocking its ability to infect normal cells. Long-term prophylaxis and passive immunization are not possible without the development of anti-globulin responses, and supplies from immune donors are limited. It is more difficult and time-consuming to raise rodent anti-viral monoclonal antibodies that can also neutralize a virus and, although it has been done, humanizing these for therapeutic use adds to the difficulties. The recombinant antibody approach is more effective at rapidly producing a wide range of different antibody specificities of human origin which could be particularly important in producing a family of anti-viral antibodies to combat frequent viral antigen changes.

Progress has been made in isolating neutralizing recombinant antibodies to respiratory syncytial virus (RSV), which is a most serious pediatric respiratory virus and also a life-threatening infection for bone marrow transplant patients. One of these antibodies has been shown to neutralize a wide range of viral isolates with a titer of less than 1 µg ml^{-1} [9] and is showing promise in combating infection in animal models simply by application through a nasal spray.

Recombinant antibodies have also been isolated from individuals infected with HIV type 1 (HIV-1), which are directed against the CD4 binding site of the gp120 envelope glycoprotein, and are effective in neutralizing the virus [10]. Antibody specificities to herpes simplex virus types 1 and 2, human cytomegalovirus, varicella zoster virus, rubella, HIV-1 and RSV have been isolated from a single combinatorial antibody gene library [11]. Thus it is apparent that virtually any anti-viral or infectious agent antibody can be isolated from an appropriate immune donor.

It is too early to know whether any of these antibodies will be effective at completely eradicating infection. However, they should certainly contribute to the reduction of viral load, thereby giving other more conventional forms of treatment, and the patient's natural immune response, the opportunity to be more efficacious.

Anti-idiotype antibodies. When the antibody raised against an antigen (e.g. a tumor-associated antigen, TAA) is itself used as an

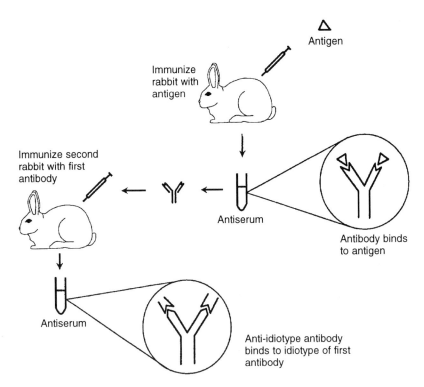

FIGURE 7.1: *Polyclonal anti-idiotype antibodies. The first rabbit is immunized with antigen and the purified antibodies from the resulting antiserum are used to immunize a second rabbit. The second antiserum contains antibodies that recognize the idiotype of the first antibodies and can therefore be a mirror image of the antigen. Anti-idiotype antibodies will also be formed to sites outside the immediate antibody combining site. See also* Figure 1.3. *Similar reactions can be elicited when a therapeutic antibody is administered to humans.*

immunogen, anti-idiotype antibodies can be produced which recognize the antigen-binding site of the original antibody (see *Figure 7.1*). Consequently, the second antibody mimics the structure of the epitope on the antigen which the first antibody recognized. If the second (anti-idiotypic antibody) is administered to patients, it will stimulate the patient's immune system to produce antibodies against it and the patient's own tumor cells will also be targeted. In addition, regulatory anti-idiotypic antibodies can activate specific T-helper cells which contribute to tumor suppression.

This approach has shown promise experimentally for many autoimmune, neoplastic and infectious diseases in illiciting radical changes in target cells, and may change the autoantibody expression in autoimmune diseases [12]. It is also beginning to show significant results clinically. For example, an anti-idiotype monoclonal antibody which is a mirror image of a melanoma-associated antigen (MAA) elicited humoral anti-MAA immunity in about 60% of patients with malignant melanoma, and was associated with survival prolongation and absence of major side effects in spite of repeated administrations [13].

All antibodies whether foreign, humanized or human recombinants will, on repeated administration, elicit the formation of an anti-idiotype response which could reduce the effectiveness of treatment. The only way of avoiding this is to administer antibodies with different idiotypes if they are available.

7.2.3 Humanized and CDR-grafted antibodies

Humanized antibodies have the effect of reducing or delaying the immunogenicity of the foreign antibody by replacing the constant regions of the rodent antibody with the human equivalent. This is taken a stage further when only the antigen-binding CDR loops of the rodent antibody are grafted into the human framework. Most of the new therapeutic antibodies to be approved over the next few years are likely to be of this type since there are many well-characterized rodent monoclonal antibodies of appropriate specificity already available. However, there will always be a likelihood of stimulating anti-idiotype antibodies even with fully humanized or recombinant antibodies.

The first example of a fully humanized CDR-grafted antibody of therapeutic potential was CAMPATH-1H [14]. This was derived from a series of CAMPATH (a trade mark of the Wellcome Foundation Ltd) rat monoclonal antibodies specific for the differentiation antigen CDw52, which is strongly expressed on virtually all human lymphocytes and monocytes but not on other blood cells, including hemopoietic stem cells. These antibodies are, therefore, potentially useful reagents for the treatment of various lymphoid malignancies, the control of GVHD in bone-marrow transplantation, and the prevention of bone-marrow and organ rejection. An IgG2b isotype was isolated by class switching from the original IgG2a-secreting hybridoma. This antibody was effective at depleting lymphocytes *in vivo* but elicited an anti-globulin response. In order to avoid this, the antibody was humanized by grafting the CDR loops on to a human IgG1 isotype which was the most effective at binding complement and T-cell activation. Initially, the affinity of the reshaped antibody was unacceptable but this was improved, with the aid of computer modeling, by changing one residue in the human framework (serine

27) to that found in the original rat framework (phenylalanine). Clinical results have been very encouraging, with none of eight rheumatoid arthritis patients mounting an anti-globulin response after the first treatment with the humanized antibody compared with 11 out of 14 treated with the original rat IgG2b antibody. Repeated treatment with the new antibody in four patients resulted in an anti-idiotype response in three, but this was not severe enough to prevent a second remission. Humanizing antibodies to reduce the HAMA response will not necessarily reduce the immunogenicity of an immunotoxin, since the toxin itself may be immunogenic (see Section 7.3.2, p. 113).

7.2.4 Recombinant antibodies

Although it is too soon since their development to have substantial clinical evidence of the benefits of recombinant antibodies, one can predict a number of potential advantages that they are likely to have over other antibodies. Human recombinant antibodies will continue to elicit anti-idiotype responses *in vivo* as do fully humanized antibodies. However, since recombinant antibodies can be obtained even without immunization, a whole new range of specificities can be obtained to traditionally poor immunogens and self-antigens, the latter being of particular interest to the study and treatment of autoimmune diseases. Another advantage is that, once a large library has been constructed, many different specificities can be selected very quickly and manipulated genetically *in vitro* to give the desired properties. This facility might have very important implications in treating viral infections that are subject to frequent changes (see also Section 7.2.2, p. 108).

7.3 Effector functions

7.3.1 Unlabeled antibodies

Rodent antibodies evolved to be most effective *in vivo* in rodents so they are not necessarily as effective as human antibodies are in humans. *In vitro* studies typically use human target cells, rat or mouse monoclonal antibodies, and rabbit or guinea-pig complement. By manipulating these components, good cell lysis can be achieved *in vitro* but cannot be used to predict the effects *in vivo* in humans. Whereas it has been known for many years which human antibody isotypes are the most effective, it was not until recombinant antibodies became available that more detailed and accurate analysis could be carried out. By studying the effects of matched sets of antibodies with identical specificity and affinity but different isotypes, variations in responses due to the variable regions could be eliminated. One of these sets was constructed from a CAMPATH-1 antibody (anti-CDw52; see Section 7.2.3) and the four human IgG subclasses, which were assayed for their ability to lyse human lymphocytes by autologous complement or by antibody-dependent cell-mediated cytotoxicity (ADCC) using activated autologous effector cells. IgG1 was clearly the best in both assays. IgG3 was also effective in both assays but IgG2 was weakly positive at binding complement and very poor at ADCC, and IgG4 did not work at all. Therefore, IgG1 antibodies are likely to be the best candidates for harnessing autologous immune systems but IgG4 might be useful, for example, in imaging or blocking applications where effector functions would complicate therapy. The ADCC response has since been shown to be more complex and variable between individuals. Some individuals respond by ADCC even with IgG4.

Experience with the original CAMPATH monoclonal antibodies showed that rat IgM, IgG1, IgG2a, IgG2b and IgG2c are all capable of lysing human lymphocytes efficiently by binding human complement. IgG2b, but not the other isotypes, is also effective in ADCC with human effector cells. More detailed molecular analysis of the antibody sequences responsible for these differences in effector function have focused on the CH2 domain and the lower hinge region, and demonstrate an important role for carbohydrate side chains. These studies are reviewed in refs 15 and 16.

7.3.2 Labeled antibodies

Radioisotopes. Radiolabeled antibodies can be used therapeutically either to aid surgeons in locating tumors and metastases prior to and during surgery or to deliver lethal doses of radioactivity directly to cancer cells, thereby reducing adverse effects on normal tissue [17]. The first clinical studies with radiolabeled antibodies began in the early 1980s and since then there have been some significant successes in terms of complete or partial treatment of disease, although the full advantages of this treatment have yet to be perfected.

There are various methods of radiolabeling antibodies depending on the choice of isotope (see Section 5.2.1). ^{125}I-labeled monoclonal antibodies can be administered to patients prior to surgery, and the γ radiation emanating from the tumor sites followed during surgery with a hand-held monitor. This method can locate masses as small as 0.1 g. Ideally, the labeled antibody should be taken up rapidly by the tumor with minimal binding to normal tissue. This method has been used effectively with antibodies directed at CEA to locate colorectal carcinoma. Technetium-99m is also commonly used for imaging since it is cheap, safe, readily available and is commonly used in standard nuclear medicine procedures (see *Figure 7.2*). [131]Iodine is particularly useful since it emits γ and β radiation. The γ emission can be visualized using a γ camera to assess the relative distribution between tumor and normal tissues, and enables the concentration of radioactivity in these tissues to be estimated. By repeating this visualization, assessment of cumulative dose and clearance rates from individual tissues can be made to aid optimization of treatment. Having established the best conditions, the same isotope can then be used for cell killing through its β emission.

The radiation energy from α- and β-emitting radioisotopes has a range of between one and 40 cell diameters. The shorter the range the less effective the dose will be at killing surrounding tissue, although this degree of localization might have advantages for treating some tumor types. The longer range β emitters (i.e. [131]Iodine and [90]Yttrium) can be useful where there is heterogeneity of antigen expression and difficulties of penetrating dense tissue. The uptake of radiolabeled antibody is directly proportional to the antigen density at the cell surface; therefore, any cell with low or no target expression would not receive a lethal dose of isotope unless it was in the direct path of radiation from the isotope. Any cancer cells surviving could lead to the proliferation of resistant or mutant cells.

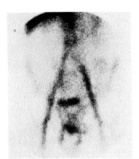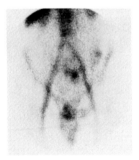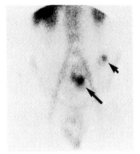

FIGURE 7.2:   *Anterior images, taken 5 min (left), 7 h (center) and 22 h (right) after injection of 600 MBq of Technetium-99m-labeled PR1A3 antibody in a patient with a mucinous adenocarcinoma of the sigmoid colon. An area of uptake, increasing with time, can be observed in the primary tumor in the centre of the image (long arrow) and in a second site, later shown to be a second adenocarcinoma, medial to the left iliac crest (short arrow). Reproduced from Granowska et al. (1993)* Eur. J. Nucl. Med. **20**, *690, Figure 3a with permission from Springer-Verlag.*

There can be problems with administering antibodies pre-labeled with radioisotopes as the radioactivity can affect nontarget cells, especially bone marrow, before the target is reached. This is known as the 'bystander effect'. It can be minimized by allowing the anti-tumor antibody (pre-conjugated with streptavidin) to reach its target first, followed by radiolabeled biotin. The radiolabel reaches its target faster and is cleared more rapidly from the body with reduced bystander toxicity. The bystander effect can also be used to advantage within a tumor to kill neighboring cells that do not express the target antigen. Clearly, the tumor size, antigen expression, isotope and concentrations all play a part in the response.

In radioimmunotherapy, doses are given continuously over a few days but are much lower than in external beam therapy being approximately 15–50 cGy h^{-1} to effect cell killing. There appears to be a strong correlation between radiation dose and response but high doses of radioactivity are also likely to damage bone marrow. It is, therefore, very important to maximize the accessibility of the tumor and the antibody concentration ratio of tumor to normal tissue. These factors can be improved by administering the radiolabel locally (i.e. directly into a body cavity or intrathecally for neural tumors) to minimize systemic toxicity. Antibody fragments penetrate more rapidly into poorly vascularized parts of tumors and clear more rapidly, but the tumor to normal ratios are improved. Higher affinity antibodies are particularly important in this respect. Enhanced clearance from normal tissues can also be achieved by giving a second antibody directed against the first anti-tumor antibody, without reducing the tumor concentration.

Immunotoxins. The types of toxin used as immunotoxins are derived from bacteria or plants, and all inhibit protein synthesis; thus, they are toxic to both dormant and actively dividing cells. The commonest plant toxins are ricin and abrin both of which are made up of two disulfide-linked polypeptides chains, A and B. The B chain is normally responsible for binding of the toxin to galactose-containing glycoproteins and glycolipids on the surface of all cell types. The toxin is then endocytosed and routed to the trans-Golgi network, where the A chain translocates to the cytosol. The A chain kills the cell by enzymatically removing an adenine residue from the 60S ribosomal subunit that binds to elongation factor 2 (EF-2) during protein synthesis. A form of diptheria toxin and *Pseudomonas* exotoxin act in a similar way in that one chain binds to a cell surface receptor and is internalized, and the second chain effects cell killing by preventing EF-2 from participating in protein synthesis. To construct an immunotoxin, it is necessary to block or inactivate or remove altogether the B chain and replace it with the specific antibody. The linker between the two, however, must be stable extracellularly but, nevertheless, allow for the toxic chain to be released in the cytosol. Recombinant immunotoxins made up of the genes of truncated diptheria toxin or *Pseudomonas* exotoxin with antibody-binding fragments have been formed which produce very stable reagents.

Toxic side effects of these immunotoxins include hepatic damage and the vascular leak syndrome (VLS). Liver toxicity from ricin A chain (RTA)-based toxins is due to the binding of mannose- and fucose-containing oligosaccharides to liver cells, resulting in rapid clearance and damage. This can be overcome by deglycosylating RTA or by using a recombinant RTA expressed in nonglycosylating cells (i.e. bacteria). The symptoms of VLS are extravasation of fluids from blood vessels into peripheral tissues causing edema, occasionally including potentially fatal pulmonary edema and also myalgia. VLS can be minimized by dose optimization.

Calicheamicins are potent DNA damaging agents of the ene-diyne class of antibiotics. They bind to the minor groove of DNA resulting in cleavage and, therefore, cell death. Their potential advantage over other immunotoxins is that they are not expected to be immunogenic. Humanized (CDR-grafted) antibody conjugates of N-acetylcalicheamicin-γ-1^1 are being developed for the treatment of ovarian and breast cancer, and acute myelogenous leukemia. The antibody targets are polymorphic epithelial mucin antigen, which is abundant on carcinomas of epithelial origin such as ovarian and breast cancer, and CD33, a cell surface glycoprotein present on normal and leukemic myeloid colony-forming cells but not on normal pluripotent hematopoietic stem cells and other normal cells. *In vitro* tests with these conjugates have highlighted the need for careful consideration of the conjugation chemistry in order to prevent impairment of both antibody binding and toxicity [18].

Lymphoid malignancies and normal lymphocytes are particularly responsive to immunotoxins because they are easily accessible. In contrast, immunotoxins are not as effective at targeting large solid tumors with poor blood supply and high interstitial pressure. This limited success of toxin conjugates (and similarly of radioisotope conjugates) is also partly due to the heterogeneity of antigen expression on tumor cells, such that even if successful killing of a particular antigen-expressing cell population was achieved, antigen deficient mutants would remain and proliferate. In such situations, a combination of several different immunotoxins and conventional cytotoxic drugs may have a synergistic anti-tumor effect. The use of immunotoxins for therapy is reviewed in ref. 19.

Cytokines. The therapeutic use of fusion proteins to combine the targeting ability of antibodies with the multifunctional activities of cytokines has been attempted successfully in animal models. Severe combined immunodeficiency (SCID) mice are a mutant strain that lack T and B lymphocytes and therefore cannot make immuno-globulin. In one example, these mice were induced to grow hepatic metastases of a human neuroblastoma. They were then given a fusion protein of a human/mouse chimeric anti-ganglioside (GD-2) monoclonal antibody co-expressed with a recombinant IL-2, together with human lymphokine-activated killer (LAK) cells. The combined treatment caused tumor cell lysis which suppressed the growth and dissemination of metastases resulting in prolonged survival of the mice. This treatment was as effective as IL-2 alone but the required dose of IL-2 was much less, thereby reducing its toxicity [20, 21].

Antibody-directed enzyme prodrug therapy. Antibody-directed enzyme prodrug therapy (ADEPT) makes use of enzyme-conjugated antibodies directed against tumor-associated antigens. When this complex is allowed to bind to tumors and clear from nontumor sites, a prodrug is administered which is converted to a cytotoxic drug by the enzyme at the tumor site. Unlike DNA toxins, in this system it is not necessary for the toxin to be internalized. The enzyme continues to turn over prodrug and the toxin can affect nonantigen-expressing cells in the immediate vicinity. There are several points in the ADEPT system which can be modified to obtain the best therapeutic results. These include the choice of target epitope to enhance tumor specificity and penetration, enhanced clearance of the antibody–enzyme complex from nontarget tissues to maximize target-to-nontarget ratios (assisted perhaps by administration of a second antibody), favorable enzyme kinetics, low toxicity of prodrug, and high toxicity but short half-life of active drug.

If the enzyme is of nonmammalian origin the possibility of nonspecific prodrug activation is minimized, although there is a risk of an immunogenic reaction to this and to antibodies of murine origin, especially with repeated treatments. Immunosuppression with cyclosporin A, however, can reduce this reaction. A prodrug that best meets the requirements of ADEPT is di-iodo phenol mustard. The first clinical trial to use this system [22] was in colorectal cancer and made use of a F(ab')$_2$ fragment specific for CEA and a bacterial carboxypeptidase (CPG2) which catalyzes the hydrolytic cleavage of reduced and nonreduced folates. The complex was activated by the mono-mesyl benzoic acid mustard drug CMDA (4-[(chloroethyl)(2-methyloxy)ethyl]amino benzoyl-L-glutamic acid). The results showed that this approach is feasible and essentially nontoxic, although the therapeutic benefit was limited due to the advanced disease of the recipients.

7.4 Regulatory issues

Every potential therapeutic drug has to be thoroughly tested in the laboratory before it can proceed to clinical trials and full approval by the regulatory authorities. Pre-clinical tests include full characterization of the antibody's specificity and cross-reactivity with other cells and tissues in different assay systems, including immunohistochemistry. All antibody-producing cell lines must go through karyology and isoenzyme analysis to ensure that any cell line is not contaminated with another. This is a theoretical problem and not one that has presented a risk in practice. Mycoplasma infection of a cell line removes it from consideration as a therapeutic agent. Approximately 4% of potentially useful lines tested are infected, albeit at low levels. The source is usually other cell lines in the production laboratory so the need for thorough routine testing of cultures during production cannot be emphasized too strongly. Viral contaminants are tested for *in vitro* and *in vivo* using the mouse antibody production test, and have been shown to be of mouse origin or from bovine serum. Endogenous retroviral contaminants (measured by electron microscopy) are a great concern. Infectious mouse retroviruses are common (~30% of cases) in mouse hybridomas but there is no evidence of detrimental clinical side effects from these in humans [23]. Human retroviruses have not been found in human cell lines or mouse–human hybridomas.

The efficacy and toxicity (both direct and indirect) of the final antibody preparation and its conjugate must be demonstrated *in vitro* and in relevant experimental models including primates. In addition, there must be rigorous quality control checks at all stages in the production of the antibody and its conjugates to ensure that scale-up procedures do not affect the product adversely and that contamination with extraneous factors is avoided.

Once these tests are complete and approved by regulatory authorities, Phase I clinical trials can proceed which involve treating a small number of patients with advanced disease in order to establish the maximally tolerated dose. Side effects, pharmacokinetics and immunogenicity are monitored but cannot be modified until the trial is over. Judging the efficacy of a new therapeutic antibody is complicated by the advanced stage of disease and the effects of previous treatments. Remodeling and retesting by scientists is often required at this stage before progression to further clinical trials.

Phase II trials involve administering a safe dose to patients with less advanced disease. Complete or partial remissions should occur in 20–40% of patients before the drug can proceed to the next stage. Phase III trials are multicenter trials of several hundred patients in comparison with matched controls receiving placebos or alternative treatments. All these trials can take years to produce a satisfactory result and many potentially useful immunotherapeutic reagents have failed to reach final approval.

7.5 Future prospects

It is clear that many of the drawbacks of using polyclonal and rodent monoclonal antibodies in human therapy (e.g. poor specificity and anti-globulin responses) can now be overcome with the advent of genetic manipulation techniques. The factors influencing effective therapy are many and multidisciplinary but, with each clinical trial, more is learnt about the best target antigens for a particular therapeutic purpose, the effector mechanisms, both natural and toxic, dosage regimes, and methods to avoid side effects. Technological developments are progressing so rapidly that they are overtaking the relatively slow process of clinical trials. Natural antibody responses are polyclonal and harness different immune mechanisms so this might suggest that cocktails of antibodies and effectors, and also other therapeutic treatments such as surgery and chemotherapy, should be considered to obtain the full benefit. There can be no doubt that, over the next few years, many new therapeutic antibodies will be approved, bringing with them significant prognostic improvements for many life-threatening diseases. By the turn of the century, the magic bullet prophecy should be a reliable if not essential part of a clinician's armoury.

References

1. Himmelweit, F. (ed.) (1960) *The Collected Papers of Paul Ehrlich,* Vol. 3. Pergamon, New York.
2. Blanchette, V.S., Kirby, M.A. and Turner, C. (1992) *Semin. Hematol.,* **29,** 72.
3. Wordell, C.J. (1991) *DICP,* **25,** 805.
4. Lane, D. and Koprowski, H. (1982) *Nature,* **296,** 200.
5. Ghosh, S. and Campbell, A.M. (1986) *Immunol. Today,* **7,** 217.
6. Isaacs, J.D. (1990) *Semin. Immunol.,* **2,** 449.
7. Queen, C., Schneider, W.P. and Waldmann, T.A. (1993) in *Protein Engineering of Antibody Molecules for Prophylactic and Therapeutic Applications in Man* (M. Clark, ed.). Academic Titles, Nottingham, p. 159.

8. Ortho Multi Centre Study Group. (1985) *N. Engl. J. Med.*, **313**, 337.
9. Barbas, C.F. III, Crowe, J.E., Cababa, D. Jones, T.M., Zebedee, S.L. Murphy, B.R. Chanock, R.M. and Burton, D.R. (1992) *Proc. Natl Acad. Sci. USA*, **89**, 10164.
10. Roben, P., Moore, J.P., Thali, M., Sodroski, J., Barbas, C.F. III and Burton, D.R. (1994) *J. Virol.*, **68**, 4821.
11. Williamson, R.A., Burioni, R., Sanna, P.P., Partridge, L.J., Barbas, C.F., III and Burton, D.R. (1994) *Proc. Natl Acad. Sci. USA*, **90**, 4141.
12. Shoenfeld, Y., Amital, H., Ferrone, S. and Kennedy, R.C. (1994) *Int. Arch. Allergy Immunol.*, **105**, 211.
13. Ferrone, S., Chen, Z.J., Liu, C.C., Hirai, S., Kageshita, T. and Mittelman, A. (1993) *Pharmacol. Ther.*, **57**, 259.
14. Reichmann, L., Clark, M. and Waldmann, H. (1988) *Nature*, **332**, 323.
15. Morrison, S.L., Canfield, S.M. and Tao, M-H. (1993) in *Protein Engineering of Antibody Molecules for Prophylactic and Therapeutic Applications in Man* (M. Clark, ed.). Academic Titles, Nottingham, p. 101.
16. Jefferis, R. and Lund, J. (1993) in *Protein Engineering of Antibody Molecules for Prophylactic and Therapeutic Applications in Man* (M. Clark, ed.). Academic Titles, Nottingham, p. 116.
17. Harrington, K.J. and Epenetos, A.A. (1994) *Clinic. Oncol.*, **6**, 391.
18. Hinman, L.M., Hamann, P.R., Wallace, R., Menendez, A.T., Durr, F.E. and Upeslacis, J. (1993) *Cancer Res.*, **53**, 3336.
19. Vitetta, E.S., Thorpe, P.E. and Uhr, J.W. (1993) *Immunol. Today*, **14**, 252.
20. Gillies, S.D., Reilly, E.B., Lo, K-M. and Reisfeld, R.A. (1991) *Proc. Natl Acad. Sci. USA*, **89**, 1428.
21. Sabzevari, H., Gillies, S.D., Mueller, B.M. Pancook, J.D. and Reisfeld, R.A. (1994) *Proc. Natl Acad. Sci. USA*, **91**, 9626.
22. Bagshawe, K.D., Sharma, S.K., Springer, C.J., Antoniw, P., Boden, J.A., Rogers, G.T., Burke, P.J., Melton, R.G. and Sherwood, R.F. (1991) *Clin. Rep. Dis. Markers*, **9**, 233.
23. FDA Co-sponsored Workshop (1992) *Preclinical Safety Testing of Monoclonal Antibodies*. Bethesda, MD.

8 Other Applications of Antibodies

8.1 Immunopurification

Affinity chromatography has become extremely popular in recent years as a means of achieving the purification of a specific product from often complex mixtures. A feature of such methods is their specificity in comparison to other purification methods such as gel filtration or ion exchange chromatography.

Affinity chromatography methods rely on the use of solid-phase matrices to which an appropriate affinity ligand is coupled. The lectin concanavalin A is an example of an affinity ligand and is usually coupled to Sepharose. This lectin binds to certain sugar residues and can therefore be used to separate glycoproteins from nonglycosylated molecules. Where specific affinity ligands are required, a logical choice would be an antibody of appropriate specificity. Classically, however, the yield of specific antibody from a conventional antiserum is so low as to make affinity matrices based on such reagents of very low capacity, unless the antibody itself can be purified substantially from the antiserum prior to coupling to the matrix. Monoclonal antibody technology is able to offer the production of large amounts of pure antibody which now makes the use of immunoaffinity purification a viable method for the isolation of product from complex mixtures.

There are several possible methods of producing immunoaffinity matrices. Probably the most widely used method is to immobilize the immunoglobulin protein that constitutes the antibody to a chemically active solid-phase matrix such as agarose or Sepharose containing succinimidyl ester or cyanogen bromide groups (*Figure 8.1*). The solid-phase matrix spontaneously binds to the immunoglobulin in slightly alkaline buffer. In these two particular examples, coupling occurs via protein amino groups and it is therefore important to use a buffer solution which does not contain amine species such as Tris. Following antibody coupling, excess active groups are often blocked with either small amine-containing species (e.g. hydroxylamine) or nonspecific protein. A further important aspect of preparation is the removal of noncovalently bound material by exposing the coupled matrix to wash cycles of alternating high and low pH, or chaotropic conditions such as high salt concentration.

Clearly, the degree of purification of product depends on the specificity of the antibody and the extent to which nonspecifically bound species are co-eluted with the required product. This is reflected by the regime used to bind and subsequently elute the material of interest. Such procedures can be performed by batch techniques, true chromatographic techniques or a combination of both. The first of these methods merely involves slurrying the affinity matrix with the mixture to be resolved, washing away unbound material (e.g. using a filter funnel to retain the solid phase) and then changing the conditions in a stepwise fashion to bring about elution of the material of interest. The second method involves forming a chromatographic column from the affinity matrix and eluting the column in a similar manner to an ion-exchange column. Here, the option exists for a gradient system to be used to bring about a change in elution conditions and thus potentially increase resolution. These two methods can be effectively combined in certain situations by performing the binding in a batch mode and the selective elution in a column chromatography mode (*Figure 8.2*).

Selective elution of the material to be purified can be achieved in a variety of ways. The most widely used methods involve manipulation of pH or the introduction of chaotropic agents. Following the washing away of unbound material, the solid-phase immune complexes can be disrupted by reducing the pH to approximately 2 or by increasing it to approximately 11. The elution buffer carries the purified material

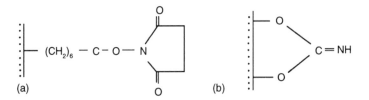

FIGURE 8.1: *Examples of activated solid phase matrices: (a) N-hydroxysuccinimide ester, (b) cyanogen bromide activated.*

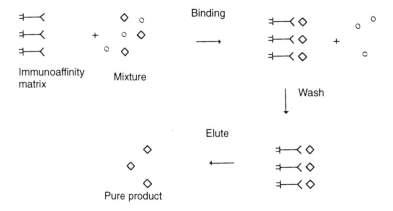

FIGURE 8.2:   *Typical procedure for immunoaffinity purification.*

away from the affinity matrix and is normally rapidly modified to less stressful conditions (e.g. by normalizing the pH to neutral) to obviate the risk of damage to the material to be isolated. In order to facilitate elution from the affinity matrix, gradual change of pH is often more effective than a sudden change, since loosely bound, and therefore potentially nonspecifically bound, material will often then be removed prior to collection of the material of interest.

A variety of species can be introduced into the elution buffer to bring about chaotropic conditions. Such agents work by disrupting bonds responsible for the integrity of the immune complexes, and include certain inorganic species such as potassium isocyanate or manganese chloride, detergents or organic solvents. Again, it is often desirable to remove the chaotropic agents from the eluent as soon as fractionation has occurred.

An important consideration in this respect is the antibody affinity. For diagnostic and therapeutic purposes, the requirement is for antibodies of high affinity. However, where the antigen has to be recovered from the immune complex, such as with immuno-purification, this may not be desirable. It can be anticipated that in such circumstances, the aggressive conditions required to disrupt the strongly bound immune complexes often result in destruction of the material to be purified. Thus, antibodies of more moderate affinity are often more suitable for these purposes. This is also important for reasons of economy where the antibody matrix is required to be recycled for further use and must thus be regenerated by removal of all antigen.

An alternative use for immunoaffinity techniques is the specific removal of a species from a mixture where the former is not required to be recovered.

8.2 Biosensors

Classical methods of biochemical analysis, including immunoassay, are based on liquid-phase chemical reaction systems. An alternative potential method is the application of some kind of solid-state measuring device or biosensor which, for example, can be used simply as an ion-selective electrode. However, the design of a biosensor based on immunochemical reactions (i.e. an immunosensor) has to be carefully considered, particularly if one wishes to be able to measure concentrations of analytes over the same range as that currently accessible using immunoassay.

A biosensor can be considered as possessing at least three domains (*Figure 8.3*): (i) the biological receptor, (ii) the transducer and (iii) the associated electronics. In an immunosensor, the biological receptor might be an antibody which must in some way be coupled to the transducer. Several possible transduction methods can be envisaged but the most commonly considered are electrochemical, optical, piezoelectric and calorimetric. The coupling of the immunochemical reaction to the transducer can be indirect or direct. In the former case, antibody binding is used to modulate a process which is detectable by the transducer. In the latter case the immunochemical reaction itself is monitored by the transducer. Two examples of the direct mode are as follows. Firstly, antibodies are immobilized on to the semiconductor gate of a field-effect transistor (*Figure 8.4*). Binding of antigens to the

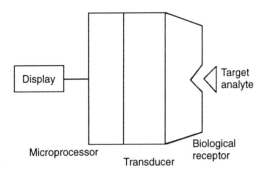

FIGURE 8.3: *Schematic representation of a general biosensor.*

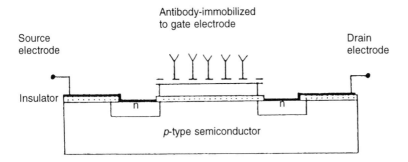

FIGURE 8.4:   *Possible configuration for an 'ImmunoFET' in which the modification of the surface charge at the gate electrode due to immune complex formation modulates the flow of current through the field-effect transistor.*

antibody layer results in a change in charge distribution and hence in the switching of the field-effect transistor. Secondly, antibodies can be immobilized on to the surface of an optical fiber. In this situation, the two most commonly exploited optical phenomena are evanescent waves and surface plasmons. An evanescent wave is produced when light passed along an optical fiber is internally reflected. This evanescent wave is the electromagnetic energy generated at the interface of the fiberoptic and the liquid. Absorption of this energy occurs if absorbing molecules are present at the interface, such that the amount of absorption is proportional to the amount of absorbing material at the interface. The binding of antigen to surface-coupled antibody can be detected in this way.

In surface plasmon resonance, a metal-coated glass optical device is used where a totally internally reflected light beam excites an induced surface electromagnetic wave or plasmon. Detectable surface plasmon resonance occurs at a particular angle of incidence of light, the latter being critically dependent on the refractive index of the medium adjacent to the metal film. Thus, changes in this layer such as might be anticipated upon immune complex formation with surface immobilized antibody can be quantified.

Several optical biosensors have also been described in which accessory reagents such as fluorescent or chemiluminescent substances have been used. These devices essentially rely on generating or quenching light at the surface of the optical fiber. These are influenced by immune complex formation with surface-immobilized antibody and detected by a photomultiplier coupled to the fiber.

8.3 Catalytic antibodies

Classically, antibodies are recognized as being functional simply in terms of their ability to bind antigen and hence act as an adapter molecule to bring into play the killing mechanisms of the immune response. The antibody binding itself does not bring about any chemical change in the antigen. This is in contrast to the similar binding of a substrate molecule to an enzyme which results in the catalytic, chemical modification of the substrate to yield an appropriate product.

The potential use of antibody binding as a means of catalyzing chemical reactions was first demonstrated by Schultz and co-workers [1] and Lerner and co-workers [2], who showed that antibodies raised to tetrahedral, negatively charged phosphonate and phosphate transition state analogs could selectively catalyze the hydrolysis of carbonates and esters, respectively. For example, ester hydrolysis involves attack by a hydroxide anion at the carbonyl carbon atom of the ester moiety. This reaction results in the production of a carboxylic acid and an alcohol as cleaved products. During this reaction, the planar ester bond system passes through a tetrahedral transition state which possesses an overall negative charge. The generation of this transition state is the rate-determining step of the reaction and is associated with an activation energy (E_a). E_a is characteristic for a particular chemical reaction. If E_a is large then a substantial activation barrier must be overcome by the input of considerable external energy and, at a given temperature, reaction rate is inversely proportional to E_a. The activation energy is lower for stabilized transition states and hence such stabilization increases the rate of reaction.

The binding and stabilization of transition states by an antibody can thus potentially result in an increase in reaction rate. However, by their nature, transition states cannot be isolated easily in order to be used as immunogens for the production of antibodies. It is therefore necessary to produce immunogens using stable molecules which bear a structural resemblance to the relevant transition state.

The example given in *Figure 8.5* shows the reaction scheme for ester hydrolysis and the type of structural analog that has been successfully used to produce catalytic antibodies capable of increasing the rate of the uncatalyzed reaction.

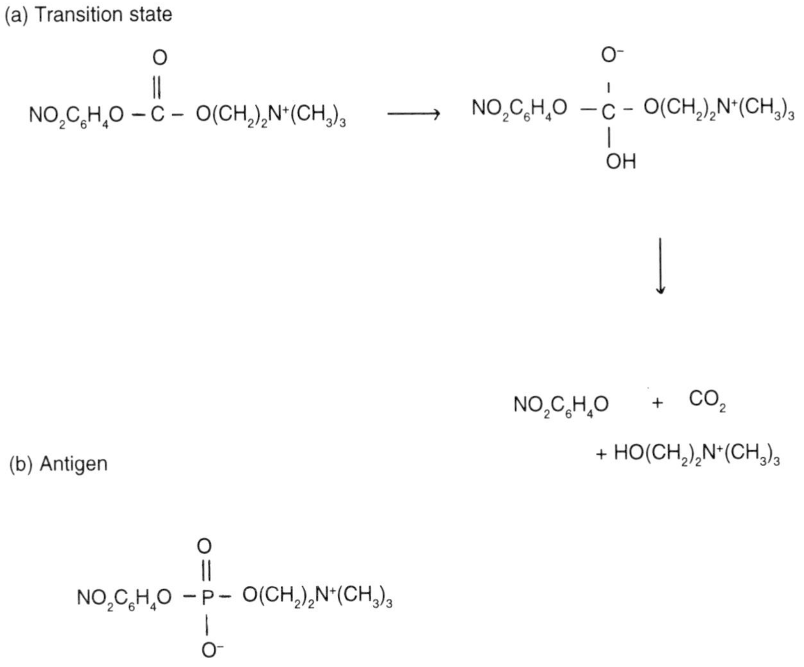

(a) Transition state

$$O \quad\quad\quad\quad\quad\quad\quad\quad O^-$$

$$NO_2C_6H_4O - \overset{\overset{\displaystyle O}{\|}}{C} - O(CH_2)_2N^+(CH_3)_3 \quad\longrightarrow\quad NO_2C_6H_4O - \overset{\overset{\displaystyle O^-}{|}}{\underset{\underset{\displaystyle OH}{|}}{C}} - O(CH_2)_2N^+(CH_3)_3$$

$$\downarrow$$

$$NO_2C_6H_4O \quad + \quad CO_2$$

$$+ HO(CH_2)_2N^+(CH_3)_3$$

(b) Antigen

$$NO_2C_6H_4O - \overset{\overset{\displaystyle O}{\|}}{\underset{\underset{\displaystyle O^-}{|}}{P}} - O(CH_2)_2N^+(CH_3)_3$$

FIGURE 8.5: *Reaction scheme for a hydrolysis reaction catalyzed by an antibody raised to the antigen (**b**). Note the similarities between (**b**) and the transition state (**a**).*

Several antibody-catalyzed reactions have been described [3], some having rate accelerations over the uncatalyzed reaction of 10^8-fold. The specificity of catalytic antibodies is such that they can exert their catalytic effect stereospecifically. This is particularly important in situations where a particular enantiomer is required either by direct stereospecific synthesis or by subsequent selective enrichment of a product mixture. It is perhaps not surprising that catalytic antibodies have been produced using monoclonal antibody technology in view of the ability of these techniques to yield large amounts of pure antibody.

8.4 Epitope mapping

The epitope specificity of monoclonal antibodies permits them to be used as reagents for the immunological characterization of the molecule to which they have been produced. Large molecules in particular can possess several distinct epitopes or, possibly, repeating epitopes. The distribution of these epitopes relative to each other and to the molecule itself can be determined using panels of monoclonal antibodies with differing epitope specificities. This characterization is often referred to as epitope mapping.

An elegant example of this approach has been described by Soos and Siddle [4] in their work on the mapping of the epitopes of human luteinizing hormone (LH). This work involved the ultimate preparation of a panel of 12 useful antibodies to LH immunogen. Initial studies of antibody specificity involved investigation of the cross-reactivity of the various antibodies with thyroid-stimulating hormone (TSH), follicle-stimulating hormone (FSH) and chorionic gonadotropin (CG). All these glycoprotein hormones possess α and β subunits. For the antibodies produced in this study, substantial cross-reactivity with TSH and FSH indicated that those particular antibodies were reacting with epitopes on the α subunit of the LH, since this subunit has the same structure in all these glycoproteins. Certain of the remaining antibodies (presumed specific for the β subunit) were found also to cross-react with CG whose β subunit has substantial structural homology with that of LH. However, other antibodies to the β subunit showed little cross-reactivity with CG, suggesting that they were specific for that portion of the LH β subunit which is known to be distinct from CG.

Competitive binding studies were used to assess the distribution of epitopes relative to each other. Here, ^{125}I-labeled LH was allowed to bind to solid-phase coupled derivatives of the various antibodies in the presence and absence of the same panel of antibodies in soluble form. Such a study demonstrated nine apparently distinct non-repeating epitopes. Partial inhibition of binding of labeled LH to solid-phase antibody by a given soluble antibody indicates spatially associated epitopes, whereas no inhibition indicates spatially separated epitopes. Lack of inhibition by a soluble antibody identical to the solid-phase antibody would be indicative of the presence of a repeating epitope on the antigen or might suggest the presence of multimeric forms of the antigen.

Such epitope mapping studies are particularly important in assessing the relative merits of pairs of antibodies used for engineering two-site immunoassays.

8.5 Fluorescence-activated cell sorting (FACS)

FACS techniques involve the use of equipment capable of resolving mixtures of cells for either analytical or preparative purposes. Cell suspensions can be passed through a nozzle such that a stream of single cells is produced. Such equipment has been used for cell cloning for monoclonal antibody production because of its ability to dispense single cells into each well of a microtiter plate. Such dispensing is nonselective. The FACS system can, however, be used to select different populations of cells. This is done by imparting an electrostatic charge to certain cells but not to others. When the stream of cells is passed through an electric field, those cells possessing a charge will be deflected whereas those which are uncharged will continue in a straight line. In this way, the charged cells can be directed into a different container to the uncharged cells.

Antibodies to cell surface antigens play an important role in enabling the FACS system to make this selection. This is most commonly done by obtaining antibodies capable of binding specifically to the cells of interest but not to the other cells within a population. These antibodies are previously tagged with fluorescent microspheres which effectively means that the cells to which the antibodies are directed become fluorescent. The stream of cells is passed through a light beam from an appropriate laser and any cells which carry antibody-coupled microspheres give rise to a fluorescent signal from a detector. It is this signal which instructs the instrument to cause a charge to be imparted to those particular cells only, and it is therefore only those cells which are deflected into the collecting vessel.

References

1. Pollack, S.J., Jacobs, J.W. and Schultz, P.G. (1986) *Science,* **234,** 1570.
2. Tramontano, A., Janda, K.D. and Lerner, R.A. (1986) *Science,* **234,** 1566.
3. Lerner, R.A., Benkovic, S.J. and Schultz, P.G. (1991) *Science,* **252,** 659.
4. Soos, M. and Siddle, K. (1983) *Clin. Chim. Acta,* **133,** 263.

Appendix A

Glossary

Adjuvant: a preparation used as a vehicle for immunogen designed to enhance the immune response following immunization.

Affinity: term used to describe the strength of binding of an antibody to an antigen or hapten.

Allotype: characteristic of an immunoglobulin molecule which is defined by allelic variation.

Antibody: an immunoglobulin molecule capable of binding to an antigen or hapten.

Antibody-directed enzyme prodrug therapy (ADEPT): a potentially less harmful means of delivering a toxic drug to its site of action since it is only converted to its toxic form at the active site.

Antigen: a molecule which induces the formation of antibody.

Ascitic fluid: serum-like fluid produced in the peritoneal cavity in response to the presence of neoplastic (i.e. hybridoma) cells.

Balb/c: an inbred strain of mice used for monoclonal antibody production.

Bispecific antibodies: chemically constructed antibodies whereby each Fab arm has a different specificity.

Class switching: change in immunoglobulin class (isotype) synthesis that occurs during differentiation of B lymphocytes.

Clone: a group of genetically identical cells.

Cloning (hybridomas): a process of cell dilution resulting in single cell isolation and subsequently formation of a clone of cells.

Combinatorial antibody gene libraries: a gene library formed from the random recombination of antibody heavy and light chain genes.

Combinatorial infection: a method for producing very large antibody gene libraries by combining two vectors containing separate heavy and light chain genes.

Competent bacteria: a transient state enabling bacteria to be transformed with DNA from external sources.

Complement: serum proteins involved in control of inflammation, activation of phagocytes and lysis of cells.

Complementarity determining region (CDR): segment of an antibody variable region containing amino acid sequences that determine antigen-binding specificity. Some or all of these regions make contact with antigen.

Cross-reactivity: reaction of an antibody with a molecule nonidentical to that which was used to elicit the immune response.

Cytokine: secreted polypeptide that affects the functions of other cells.

dAbs (single-domain antibodies): Single antigen-binding VH domains derived from the cloning of VH genes and expression of the peptide by *E. coli*.

Effector function: any biological activity of an antibody other than antigen binding.

Electroacoustic fusion: a method of electrically fusing cells by first aligning them in an ultrasonic sound wave.

Electrofusion: fusion of cells induced by an electric pulse.

Electrophoresis: a method of separating polypeptides, fragments or nucleic acids within a gel under the influence of an electric field.

Electroporation: a method of introducing molecules (i.e. DNA) into cells by subjecting them to a brief high-voltage electric pulse.

Fab (fragment antigen binding): one of two parts of an antibody molecule responsible for binding to antigen which can be separated by digestion of immunoglobulin by papain.

F(ab')$_2$: a bivalent antigen binding fragment of immunoglobulin formed by digestion with pepsin.

Fc (fragment crystallizable): part of an antibody responsible for effector function. It can be separated by digestion of immuno-globulin by papain.

Fc receptor: receptor on a variety of cell types that can bind to the Fc part of immunoglobulin.

Fusion: the combination of two cells to form a single cell (i.e. hybridoma) that contains the genetic material of both parents.

Fv (fragment variable): a fragment formed from one VH and one VL domain held together by noncovalent interactions. It contains one antigen binding site.

Hapten: small molecule which constitutes an epitope but which cannot itself elicit an immune response.

HAT medium: cell culture medium used to separate hybridomas from myeloma cells following fusion. Myeloma cells normally cannot survive in HAT medium.

Helper phage: virus with a defective origin of replication used to co-infect bacteria containing phagemid. Secretion of the phagemid package is favored.

Heterohybridomas: hybridomas derived from the fusion of lymphocytes of one species and myeloma cells of another species.

Heteromyeloma: a myeloma cell line derived from the fusion of two different myeloma cell lines from different species.

Hinge region: the flexible region of an immunoglobulin molecule that joins the two Fab arms to the Fc region.

Human anti-mouse antibody (HAMA) response: an immune response of a human recipient of therapeutic mouse monoclonal antibodies, especially as a result of repeated doses. It neutralizes the effect of the mouse antibody and causes formation of immune complexes and allergic reactions in the patient.

Humanized antibody: a rodent monoclonal antibody in which the nonantigen binding part has been genetically replaced by a human antibody

Hybrid hybridomas: hybridomas produced from the fusion of two hybridomas of different specificity or one hybridoma and lymphocytes of different specificity.

Hybridoma: a cell formed from the fusion of a myeloma cell and antibody-producing lymphocyte.

Idiotype: antigenic characteristic of an immunoglobulin variable region.

Immunoassay: method of using antibodies to detect and quantify analytes.

Immunoblotting: a process whereby antigens are transferred to paper after electrophoresis for the application of antibody probes.

Immunocytochemistry: methods for visualizing antigens *in situ*.

Immunogen: antigenic preparation administered during immunization.

Immunopurification: method of using antibodies in specific purification procedures.

Immunotoxin: an antibody chemically or genetically fused to a toxin.

***In vitro* immunization:** stimulation of lymphocytes to produce specific antibody *in vitro*.

Isotype: description of the various classes of immunoglobulin molecules.

Limiting dilution: a method of cloning hybridoma cells.

Lymphocyte: one of a group of cells involved in the immune system.

Lymphokine: humoral products of lymphocytes involved in immune system signaling.

Monoclonal: deriving from a single clone of cells.

Myeloma: neoplastic proliferation of a clone of plasma cells. Myeloma cell lines have been adapted for use as malignant parents for the production of hybridomas by cell fusion.

Naive antibody gene library: a gene library produced from nonimmunized sources.

Paratope: that part of an antibody molecule which contacts with an epitope.

Phage: virus that infects bacteria. Used for cloning DNA.

Phage display: a technique whereby a protein is displayed on the external surface of a bacteriophage enabling selection of phage containing the DNA coding for that protein.

Phagemid vector: plasmid with an origin of replication for filamentous bacteriophage. Bacteria infected with phagemids require co-infection with helper phage to produce free phage particles.

Phagocyte: a blood cell capable of engulfing foreign material.

Plasmacytoma: *see* Myeloma.

Plasmid: extrachromosomal double-stranded circular DNA found in most bacterial species, dependent on the replication machinery of the host cell. Can be used as cloning vectors for DNA.

Polyclonal: deriving from more than one clone of cells.

Polymerase chain reaction (PCR): a method of amplifying small sequences of DNA.

Polymerase chain reaction (PCR) primer: a small sequence of DNA complementary to either end of the template DNA to be amplified by PCR.

Radioimmunotherapy: a method of targeting a toxic radioisotope to antigen (i.e. on a tumor) by radiolabeling a specific antibody.

Random or site-directed mutagenesis: a method of changing a specific nucleic acid residue resulting in a change in the resulting protein.

Severe combined immunodeficient (SCID) mice: a mutant strain that lack T and B lymphocytes and therefore cannot make immunoglobulin.

Titer: semi-quantitative measure of the amount of binding activity in an antibody preparation.

Transfection: the transfer of DNA into bacteria or animal cells.

Transgenic mice: mice in which foreign DNA has been introduced into the genome such that it is transmitted to subsequent generations.

Vascular leak syndrome (VLS): toxin side effect causing extravasation of fluids from blood vessels into peripheral tissues causing edema and occasionally fatal pulmonary edema and myalgia.

Appendix B

Suppliers

This list contains the UK and USA addresses of the major suppliers of antibodies and related equipment. It is not exhaustive but should provide a useful starting point for those beginning to use antibodies for research and diagnostics. The most comprehensive list of suppliers, including national distributors worldwide, can be found in Linscotts Directory of Immunological and Biological Reagents and the MSRS Catalog.

1. Antibodies, antibody conjugates, assay kits and related equipment

Accurate Chemical & Scientific Corp., 300 Shames Drive, Westbury, NY 11590, USA. Tel. 800 645 6264; Fax 516 997 4948.

Amersham International plc, Amersham Place, Little Chalfont, Bucks HP7 9NA, UK. Tel. 0800 616928; Fax 0800 616927.
Amersham Corporation, 2636 South Clearbrook Drive, Arlington Heights, IL 60005, USA. Tel. 800 323 9750.

Becton-Dickinson UK Ltd, Between Towns Road, Cowley, Oxford OX4 3LY, UK. Tel. (0)1865 748844; Fax (0)1865 717313.
Becton-Dickinson Immunocytometry Systems, 2350 Qume Dr., San Jose, CA 95131-1807, USA. Tel. 800 223 8226; Fax 408 954 2009.

Biodesign International, 105 York St, Kennebunk Port, Maine, MA 04043, USA. Tel. 207 985 1944; Fax 207 985 6322.

Bio-Rad Laboratories Ltd, Bio-Rad House, Maylands Ave, Hemel Hempstead, Herts HP2 7TD, UK. Tel. 0800 181134; Fax (0)1442 259118.
Rio-Rad Laboratories, 2000 Alfred Nobel Dr., Hercules, CA 94547, USA. Tel. 510 741 6808; Fax 510 741 1051.

Boehringer Mannheim UK (Diagnostics and Biochemicals) Ltd, Boehringer Mannheim House, Bell Lane, Lewes, East Sussex BN7 1LG, UK. Tel. (0)1273 480444; (0)1273 480226.
Boehringer Mannheim Biochemicals, P.O. Box 50414, Indianapolis, IN 46250-0414, USA. Tel. 800 262 1640; Fax 317 576 2754.

Calbiochem, P.O. Box 12087, La Jolla, CA 92039-2087, USA. Tel. 800 854 3417; Fax 800 776 0999.
Novabiochem (UK) Ltd, 3 Heathcote Building, Highfields Science Park, University Boulevard, Nottingham NG7 2QJ, UK. Tel. (0)1602 430840; Fax (0)1602 430951.

Cambridge Bioscience, 25 Signet Court SCBC, Swann's Rd, Cambridge CB5 8LA, UK. Tel. (0)1223 316855; Fax (0)1223 60732.

Ciba-Corning Diagnostics Ltd, Colchester Rd, Halstead, Essex CO9 2DX, UK. Tel. (0)1787 474742; Fax (0)1787 475088.
Ciba-Corning Diagnostics, 63 North St, Medfield, MA 02052, USA. Tel. 800 255 3232; Fax 508 359 3599.

Dako Ltd, 16 Manor Courtyard, Hughenden Avenue, High Wycombe, Bucks HP13 5RE, UK. Tel. (0)1494 452016; Fax (0)1494 441846.
Dako Corporation, 6392 Via Real, Carpinteria, CA 93013, USA. Tel. 805 566 6655; Fax 805 566 6688.

Dynal (UK) Ltd, Station House, 26 Grove St, New Ferry, Wirral, Merseyside L62 5AZ, UK. Tel. (0)151 644 6555; Fax (0)151 645 2094.
Dynal Inc., 475 Northern Boulevard, Great Neck, NY 11021, USA. Tel. 516 829 0039; Fax 516 829 0045.

Genzyme Corp., 1 Kendall Square, Cambridge, MA 02139, USA. Tel. 800 332 1042; Fax 617 374 7300.
Genzyme Diagnostics, 50 Gibson Drive, Kings Hill, West Malling, Kent ME19 6HG, UK. Tel. 0800 373415; Fax (0)1732 220024.

GIBCO BRL – *see* Life Technologies Ltd.

Harlan Bioproducts for Science Inc., P.O. Box 29176, Indianapolis, IN 46229, USA. Tel. 317 894 7536; Fax 317 894 1840.

Harlan-Sera-Lab Ltd, Crawley Down, Sussex RH10 4FF, UK. Tel. (0)1342 716366; Fax (0)1342 717351.
US Distributor: Harlan Bioproducts for Science Inc.

ICN Biomedicals Ltd, Eagle House, Peregrine Business Park, Gomm Rd, High Wycombe, Bucks HP13 7DL, UK. (0)1494 443826; Fax (0)1494 473162.
ICN Biomedicals Diagnostics Div., 3300 Hyland Ave, Costa Mesa, CA 92626, USA. Tel. 800 854 0530; Fax 800 334 6999.
ICN Immunologicals, Research Products Division, P.O. Box 5023, Costa Mesa, CA 92626, USA. Tel. 800 854 0530; Fax 714 557 4872.

Jackson Immunoresearch Laboratories Inc., P.O. Box 9, West Grove, PA 19390, USA. Tel. 800 367 5296; Fax 215 869 0171.
UK Distributor: Stratech Scientific Ltd.

Life Sciences International (Europe) Ltd, Chadwick Rd, Astmoor, Runcorn, Cheshire WA7 1PR, UK. Tel. (0)1918 566671; Fax (0)1928 565845.

Life Technologies Ltd, P.O. Box 35, Trident House, Renfrew Rd, Paisley PA3 4EF, UK. Tel. (0)141 889 6100; Fax (0)141 887 1167.
Life Technologies Inc., 8400 Helgerman Court, Gaithersburg, MD 20877, USA. Tel. 301 840 8000.

Pharmacia Biotech, 23 Grosvenor Rd, St Albans, Herts AL1 3AW, UK. Tel. (0)1727 814000; Fax (0)1727 814001.
Pharmacia Biotech, 800 Centennial Ave, P.O. Box 1327, Piscataway, NJ 08855-1327, USA. Tel. 800 526 3593; Fax 800 329 3593.

Pierce & Warriner, 44 Upper Northgate St, Chester CH1 4EF, UK. Tel. (0)800 252185; Fax (0)1244 373212.
Pierce, 3747 North Meridian Rd, P.O. Box 117, Rockford, IL 61105, USA. Tel. 815 968 0747; Fax 800 842 5007.

Promega Corporation, 2800 Woods Hollow Rd, Madison, WI 53711, USA. Tel. 800 356 9526; Fax 608 277 2516.
Promega UK, Delta House, Enterprise Rd, Chilworth Research Centre, Southhampton SO1 7NS, UK. Tel. (0)1703 760225; Fax (0)1703 767014.

Scripps Laboratories, 11180 Roselle St, San Diego, CA 92121-1211, USA. Tel 619 546 5800; Fax 619 546 5812.
UK Distributor: Cambridge Bioscience.

Serotec Ltd, 22 Bankside, Station Field Industrial Estate, Kidlington, Oxford OX5 1JE, UK. Tel. (0)1865 379941; Fax (0)1865 373899.
US Distributor: Harlan Bioproducts for Science Inc.

Sigma Chemical Co. Ltd, Fancy Rd, Poole, Dorset BH17 7NH, UK. Tel. 0800 373731; Fax 0800 378785.
Sigma Chemical Co., P.O. Box 14508, St Louis, MO 63178, USA. Tel. 800 325 3010; Fax 800 325 5052.

Stratech Scientific Ltd, 61–63 Dudley St, Luton, Beds LU2 0NP, UK. Tel. (0)1582 481884; Fax (0)1582 481895.

The Binding Site, Institute of Research and Development, Vincent Dr., Birmingham B15 2SQ, UK. Tel. (0)121 471 4197; Fax (0)121 472 6017.
5889 Oberlin Dr. #101, San Diego, CA 92121, USA. Tel. 619 453 9177; Fax 619 453 9189.

Wellcome Diagnostics Ltd, Temple Hill, Dartford DA1 5AH, UK. Tel (0)1322 277711; Fax (0)1322 273288.
Wellcome Diagnostics Division, North Building, 3030 Cornwallis Rd, Research Triangle Park, NC 27709, USA. Tel. 800 334 9332.

Zymed Laboratories, 458 Carlton Court, South San Francisco, CA 94080, USA. Tel. 800 874 4494; Fax 415 871 4499.
UK Distributor: Cambridge Bioscience.

2. Antibody-related product databases

Linscotts Directory of Immunological and Biological Reagents, P.O. Box 55, East Grinstead, Sussex RH19 3YL, UK. Tel. (0)1342 824854.
4877 Grange Rd, Santa Rosa, CA 95404, USA. Tel. 707 544 9555.

MSRS Catalog, Aerie Corp., P.O. Box 1356, Birmingham, MI 48012, USA.

3. Antibody-related reference databases (also available on disc)

Sheffield University Biomedical Information Services (SUBIS) Bulletin on Monoclonal Antibodies, University of Sheffield, Sheffield S10 2TN, UK. Tel. (0)174 768555, ext. 6232.

CABS – Current Advances in Immunology and Infectious Diseases. Subscriptions: Elsevier Science Ltd, P.O. Box 800, Kidlington, Oxford OX5 1DX, UK. Tel. (0)1865 843000; or Elsevier Science Inc., 660 White Plains Rd, Tarrytown, NY 10591-5153, USA. Tel. 914 524 9200.

See also general computer databases such as BIDS and Medline.

4. Cell banks

American Type Culture Collection (ATCC), 12301 Parklawn Drive, Rockville, MD 20852, USA. Tel. 800 638 6597/301 881 2600; Fax 301 231 5826.

European Collection of Animal Cell Cultures (ECACC), PHLS CAMR, Porton Down, Salisbury, Wiltshire SP4 0JG, UK. Tel (0)1980 612511; Fax (0)1980 611315.

5. Custom antibody production

See antibody product databases but also try university departments.

Appendix C

Further reading

Clark. M, (ed.) (1993) *Protein Engineering of Antibody Molecules for Prophylactic and Therapeutic Applications in Man.* Academic Titles, Nottingham.

Committee for Proprietary Medicinal Products: Ad hoc Working Party on Biotechnology/Pharmacy and Working Party on Safety Medicines. (1991) *Production and Quality Control of Human Monoclonal Antibodies.* EEC Regulatory Document. *Biologicals,* **19**, 133.

Hermanson, G.T., Mallia, A.K. and Smith, P.K. (1992) *Immobilised Affinity Ligand Techniques.* Academic Press, San Diego.

Langone, J.J. and Van Vanukis, H. (eds) *Methods in Enzymology (Immunochemical Techniques),* Vols 70 (1980), 73 (1981), 74 (1981), 84 (1982), 92 (1983). Academic Press, New York.

Liddell, J.E. and Cryer, A. (1991) *A Practical Guide to Monoclonal Antibodies.* John Wiley & Sons, Chichester.

Parks, D.R. and Herzenberg, L.A. (1984) Fluorescence-activated cell sorting: theory, experimental optimization, and applications in lymphoid cell biology. *Meth. Enzymol.,* **108**, 197.

Roitt, I.M., Brostoff, J. and Male, D.K. (1989) *Immunology,* 2nd Edn. Churchill Livingstone, Edinburgh.

Tijssen, P. (1985) *Practice and Theory of Enzyme Immunoassays.* Laboratory Techniques in Biochemistry and Molecular Biology, Vol. 15 (R.H. Burdon and P.H. van Knippenberg, eds). Elsevier Science Publishers, Amsterdam.

Turner, A.P.F., Karube, I. and Wilson, G.S. (1987) *Biosensors: Fundamentals and Applications.* Oxford University Press, Oxford.

Wawrzynczak, E.J. (1995) *Antibody Therapy.* BIOS Scientific Publishers, Oxford.

Weeks, I. (1992) *Cheiluminescence Immunoassay*. Comprehensive Analytical Chemistry Series (G. Svehla, ed.). Elsevier Science Publishers, Amsterdam.

Weir D.M. (ed.) (1986) *Handbook of Experimental Immunology*. Blackwell, Oxford.

Winter, G. and Harries, W.J. (1993) Humanized antibodies. *Immunol. Today,* **14,** 243.

Winter, G., Griffiths, A.D., Hawkins, R.E. and Hoogenboom, H.R. (1994) Making antibodies by phage display technology. *Ann. Rev. Immunol.,* **12,** 433.

Zola, H. (ed.) (1994) *Monoclonal Antibodies: the Second Generation*. BIOS Scientific Publishers, Oxford.

Index

α-naphthol/pyronin, 91
Abrin, 114
Acridinium salts, 80
Acute infection, human therapy, 104
Adjuvant, 10, 131
AEC (3-amino-9-ethylcarbazole), 91
Affinity, 15–18, 60, 131
Agglutination assays, 68
Alkaline phosphatase, 77, 91–92
Allotypic variation, 5, 131
AMCA (7-amino-4-methyl-coumarin-3-acetic acid), 90
American Type Culture Collection, 28, 139
Aminopterin, 31
Animals (Scientific Procedures) Act, UK (1986), 10, 36
Anti-17-1A antibodies, 107
Anti-cancer antibodies, 106–107
Anti-carboxypeptidase (CPG2), 116
Anti-carcinoembryonic antigen (CEA) antibodies, 107, 113, 116
Anti-ganglioside antibodies, 116
Anti-globulin responses, 104–105, 110–111
Anti-idiotype antibodies, 6, 108–110
Anti-immune system antibodies, 107–108
Anti-lymphocyte globulin, 104
Anti-thymocyte globulin, 104
Anti-viral antibodies, 108
Antibody (see also Immunoglobulin), 131
  affinity, 15–18, 60
  binding capacity, 18
  biosynthesis, 3
  classes, 2
  cross-reactivity, 18–19
  -directed cellular cytotoxicity (ADCC), 112
  -directed enzyme prodrug therapy (ADEPT), 116
  diversity, 3–4
  expression, 57–59, 61
  fragmentation, 46–49
  genes, 45
  labels, 23, 89–94
  structure, 4, 46
  subclasses, 2–4
  titer, 12–15
Antigen, 9, 131
Antigenic determinant (see also Epitope), 25
Antiserum, 12
APAAP (see also Alkaline phosphatase), 96–97
Ascitic fluid, 131
  immunocytochemistry, 88
  production of, 26, 36
Autofluorescence, 100
Autoimmune cytopenias, 105
Autoimmune diseases, 105, 107
Avidin, 95–97
Azaguanine, 29

B-cell chronic lymphocytic leukemia, 105
B lymphocytes, 1–2, 27, 38–39
Bacterial infections, human therapy, 105
Balb/c mice, 26, 36. 131
BCIP/NBT, 91–92
Binding capacity, 18
Biosensors, 124–125
Biotin, 95–97
Bispecific antibodies, 45, 48–49, 131
Bone-marrow transplantation, 110
Botulism, 104
Bystander effect, 114

Calicheamicins, 115
CAMPATH antibodies, 108, 110–112
Capping, 100
Catalytic antibodies, 126–127
CD20, 107
CD22, 107
CD3, 107
CD33, 115

143

ORDERING DETAILS

Main address for orders

BIOS Scientific Publishers Ltd
9 Newtec Place, Magdalen Road,
Oxford OX4 1RE, UK
Tel: (0)1865 726286
Fax: (0)1865 246823

Australia and New Zealand
DA Information Services
648 Whitehorse Road, Mitcham, Victoria 3132, Australia
Tel: (03) 873 4411
Fax: (03) 873 5679

India
Viva Books Private Ltd
4325/3 Ansari Road, Daryaganj, New Delhi 110 002, India
Tel: 11 3283121
Fax: 11 3267224

Singapore and South East Asia
(Brunei, Hong Kong, Indonesia, Korea, Malaysia, the Philippines,
Singapore, Taiwan, and Thailand)
Toppan Company (S) PTE Ltd
38 Liu Fang Road, Jurong, Singapore 2262
Tel: (265) 6666
Fax: (261) 7875

USA and Canada
Books International Inc
PO Box 605, Herndon, VA 22070, USA
Tel: (703) 435 7064
Fax: (703) 689 0660

Payment can be made by cheque or credit card (Visa/Mastercard, quoting number and expiry date). Alternatively, a pro forma invoice can be sent.

Prepaid orders must include £2.50/US$5.00 to cover postage and packing for one item. Prepaid orders for two or more books are delivered postage free.